
RADIATION PROCESSES IN ASTROPHYSICS

RADIATION PROCESSES IN ASTROPHYSICS

Wallace H. Tucker

The MIT Press
Cambridge, Massachusetts, and London, England

PUBLISHER'S NOTE

This format is intended to reduce the cost of publishing certain works in book form and to shorten the gap between editorial preparation and final publication. The time and expense of detailed editing and composition in print have been avoided by photographing the text of this book directly from the author's typescript.

Printed in the United States of America

Library of Congress Cataloging in Publication Data

Tucker, Wallace H
Radiation processes in astrophysics.

Includes bibliographies.
1. Astrophysics. 2. Radiation. I. Title.
QB461.T8 523.01 75-29236
ISBN 0-262-20021-X
ISBN 0-262-70010-7 pbk.

This book is intended to serve as a guide to understanding radiation processes in astrophysics and as a textbook for a senior or graduate level course in astronomy and astrophysics. No attempt has been made at an original or particularly elegant treatment of the subject matter. Rather the aim has been to pull together in a reasonably coherent manner some of the results of radiation theory which are commonly used in astronomical applications. In the first two chapters I present, often without proof, the basic results of the classical and quantum theories of radiation. In the remaining chapters I discuss applications of this basic theory to particular physical processes. Additional applications are given in the problems at the end of each chapter.

I am grateful to George Blumenthal, Stan Olbert, Bruno Rossi and Alex Dalgarno for useful suggestions, to the MIT Press for patiently waiting for a manuscript that was more than two years overdue, and to the typist, Mrs. Sheila Toomey, for doing an excellent job on a difficult manuscript.

RADIATION PROCESSES IN ASTROPHYSICS

Chapter 1

BASIC FORMULAS FOR CLASSICAL RADIATION PROCESSES

In many ways the classical description of a given radiation process is the easiest one to visualize. Usefulness rather than rigor is the goal of this book, so the classical or semi-classical descriptions will be used whenever possible.

In general, classical physics applies when the de Broglie wavelength of the radiating particle is small compared to the typical dimensions of the problem, i.e., when

$$h/p \ll r$$

here h is Planck's constant, p is the momentum of the radiating particle and r is a typical dimension of the system. Stated another way, the uncertainty in the position of the particle must be much less than the characteristic dimensions of the problem. The dimension r may refer to the effective radius of the interaction, or to the wavelength of the radiation. The classical domain can also be defined in terms of the energy W of the radiating particle and the frequency ν of the emitted radiation. Since $W \simeq pv$ and $\nu \simeq v/r$, where v is the velocity of the particle, we have

$$h\nu \ll W$$

This condition states that a classical particle cannot convert a significant amount of its energy into one photon, or alternatively, that the classical approximation holds only for transitions in which the relative change in the principle quantum number is small.

In this chapter the basic formulas needed to calculate classical radiation processes are summarized.

1.1 The Electromagnetic Field Equations

The classical theory of radiation is based on Maxwell's theory of the electromagnetic field. For a given distribution of charge density ρ and current density $\vec{j}$ the field is determined by Maxwell's equations:

$$\text{curl } \vec{E} = -\frac{1}{c}\frac{\partial \vec{B}}{\partial t} \qquad (1\text{-}1)$$

$$\text{curl } \vec{B} = \frac{4\pi}{c}\vec{j} + \frac{1}{c}\frac{\partial \vec{E}}{\partial t} \qquad (1\text{-}2)$$

$$\text{div } \vec{E} = 4\pi\rho \qquad (1\text{-}3)$$

$$\text{div } \vec{B} = 0 \qquad (1\text{-}4)$$

(CGS units will be used unless specifically stated otherwise.) Here $\vec{E}$ and $\vec{B}$ denote the electric and magnetic fields and c denotes the speed of light in vacuum. From these equations it follows that the electric charge is conserved, i.e., that it satisfies a continuity equation:

$$\text{div } \vec{j} + \frac{\partial \rho}{\partial t} = 0 \qquad (1\text{-}5)$$

The motion of the particles is described by Newton's second law

$$d\vec{p}/dt = \vec{F} \qquad (1\text{-}6)$$

where $\vec{p}$ is the momentum of the particle and $\vec{F}$ the Lorentz force, which for particles of charge e and velocity $\vec{v}$ is given by

$$\vec{F} = e(\vec{E} + \frac{\vec{v} \times \vec{B}}{c}) \tag{1-7}$$

The fields which are to be used in the Lorentz equation are the external fields as well as the fields produced by the charge itself. This self-produced field will also affect the motion of the particle. The reaction of the field in general has only a small influence on the short-term motion of the particle, so that to a first approximation, the external fields only may be used in (1-6).

The rate of change of the kinetic energy W_k of a charge in an electromagnetic field is given by

$$dW_k/dt = \vec{v} \cdot d\vec{p}/dt = e\ \vec{v} \cdot \vec{E} \tag{1-8}$$

The magnetic field does not enter into this equation since the force which the magnetic field exerts on the charge is always perpendicular to its velocity, and hence does no work on it. The rate of increase of energy of all the particles in a unit volume is found by summing (1-8) over all particles in that volume. The result is

$$dU_p/dt = \vec{j} \cdot \vec{E} \tag{1-9}$$

where U_p is the kinetic energy density of the particles. In a given volume it changes with time because of changes in the energy of the particles, and because of the flow of particles out of the volume. Thus

$$dU_p/dt = \partial U_p/\partial t + \text{div}\,\vec{Q} = \vec{j}\cdot\vec{E} \quad (1\text{-}10)$$

where $\vec{Q}$ is the energy flux density vector. For a particle distribution function $f(\vec{r},\vec{p},t)$, $\vec{Q}$ is given by

$$\vec{Q} = \int W\,\vec{v}\,f(\vec{r},\vec{p},t)\,d^3\,V \quad (1\text{-}11)$$

Through the use of Maxwell's equations and some vector algebra, the scalar product $\vec{j}\cdot\vec{E}$ can be written in the form

$$j\cdot E = -\frac{\partial}{\partial t}\frac{(E^2 + B^2)}{8\pi} - \text{div}\frac{(c\vec{E} \times \vec{B})}{4\pi} \quad (1\text{-}12)$$

Equation (1-10 can now be rewritten as

$$\frac{\partial}{\partial t}\left\{U_p + \frac{(E^2 + B^2)}{8\pi}\right\} + \text{div}\left\{\vec{Q} + \frac{c\vec{E} \times \vec{B}}{4\pi}\right\} = 0 \quad (1\text{-}13)$$

If we intergrate (1-13) over a volume V, and use Gauss' theorem to express the divergence term as a surface integral, then we obtain

$$\frac{\partial}{\partial t}\int\left\{U_p + \frac{(E^2 + B^2)}{8\pi}\right\}dV = -\int(\vec{Q} + \vec{S})\circ d\vec{A} \quad (1\text{-}14)$$

where

$$\vec{S} = (c\vec{E} \times \vec{B})/4\pi \quad (1\text{-}15)$$

is called the Poynting vector.
For a closed system in which there are no fields at the boundary and no heat transfer

across the boundary, the surface integral vanishes, and the quantity on the left hand side of equation (1-14) is conserved. The first term in the brackets represents the kinetic energy density, so the second term must be the energy density of the electromagnetic field:

$$U_{em} = (E^2 + B^2)/8\pi \tag{1-16}$$

In general the surface integral will not vanish; the first term gives the rate at which heat flows into or out of the volume. The second term therefore gives the flux of electromagnetic energy across the boundary, and the Poynting vector $\vec{S}$ is the amount of electromagnetic energy passing through a unit surface area in a unit time.

The equation for the rate of change of linear momentum can be expressed in a form which is similar to equation (1-14). From equations (1-6) and (1-7) it follows that the rate of change of the momentum density of the particles is (assuming that the particle pressure is negligible):

$$d\vec{\Pi}_p/dt = \rho\vec{E} + (\vec{j} \times \vec{B})/c \tag{1-17}$$

Using Maxwell's equations to eliminate ρ and $\vec{j}$, we may write equation (1-17) in the form (see Jackson, 1962)

$$\frac{\partial}{\partial t} \int \{\vec{\Pi}_p + \frac{\vec{E} \times \vec{B}}{4\pi c}\} \, dV = \int \overleftrightarrow{T} \cdot d\vec{A} \tag{1-18}$$

Here the tensor $\overleftrightarrow{T}$, called Maxwell's stress tensor, is in dyadic notation

$$\overleftrightarrow{T} = \{\vec{E}\,\vec{E} + \vec{B}\,\vec{B} - \frac{1}{2}\overleftrightarrow{I}\,(E^2 + B^2)\}/4\pi \qquad (1\text{-}19)$$

In tensor notation, the elements are given by

$$T_{ij} = \{E_i\,E_j + B_i\,B_j - \frac{1}{2}\,\delta_{ij}(E^2 + B^2)\}/4\pi \qquad (1\text{-}20)$$

For a closed system the integral on the right hand side of (1-17) vanishes, so the second term on the left is to be identified with the electromagnetic momentum. The electromagnetic momentum density, $\vec{\Pi}_{em}$, is therefore

$$\vec{\Pi}_{em} = (\vec{E} \times \vec{B})/4\pi c \qquad (1\text{-}21)$$

1.2 Constant Electromagnetic Fields

In the special case when $\vec{B} = 0$ and all time derivatives vanish, Maxwell's equations become

$$\vec{\nabla} \cdot \vec{E} = 4\pi\rho \qquad (1\text{-}22)$$

$$\vec{\nabla} \times \vec{E} = 0 \qquad (1\text{-}23)$$

Equation (1-23) shows that the electric field E can be expressed in terms of a scalar potential

$$\vec{E} = -\vec{\nabla}\varphi \qquad (1\text{-}24)$$

Equation (1-22) then becomes

$$\nabla^2\varphi = -4\pi\rho \qquad (1\text{-}25)$$

Solving this equation yields the potential

$\varphi(\vec{R}_o)$ at $\vec{R}_o$ due to a charge distribution ρ:

$$\varphi(\vec{R}_o) = \int \frac{\rho(\vec{R}')\, dV'}{|\vec{R}_o - \vec{R}'|} \tag{1-26}$$

For a system of point charges e_i located at $\vec{R}_i$ the potential at $\vec{R}_o$ is

$$\varphi(\vec{R}_o) = \Sigma_i\, e_i \,/\, |\vec{R}_o - \vec{R}_i| \tag{1-27}$$

For the case when $\vec{E} = 0$ and all time derivatives vanish, the magnetic field must satisfy the equations

$$\vec{\nabla} \cdot \vec{B} = 0 \tag{1-28}$$

$$\vec{\nabla} \times \vec{B} = 4\pi\vec{j}/c \tag{1-29}$$

If we introduce the <u>vector potential</u> $\vec{A}$ which satisfies the conditions

$$\vec{\nabla} \times \vec{A} = \vec{B} \tag{1-30}$$

$$\vec{\nabla} \circ \vec{A} = 0 \tag{1-31}$$

then the equation (1-29) becomes

$$\nabla^2 \vec{A} = -4\pi\vec{j}/c \tag{1-32}$$

In analogy with (1-26) the desired solution for $\vec{A}$ is

$$\vec{A}(\vec{R}_o) = \frac{1}{c} \int \frac{\vec{j}(\vec{R}')\, dV'}{|\vec{R}_o - \vec{R}'|} \tag{1-33}$$

For a system of point charges e_i with

velocities $\vec{v}_i$

$$\vec{A}\,(\vec{R}_o) = \frac{1}{c}\,\Sigma_i\; e\vec{v}_i/(\vec{R}_o - \vec{R}_i) \qquad (1\text{-}34)$$

1.3 The Dipole and Higher Moments

Consider the field produced by a system of charges at distances large compared with the dimensions of the system.

For $R_o >> R_i$ equation (1-27) becomes

$$\varphi = \frac{\Sigma\, e_i}{R_o} - \vec{d} \cdot \vec{\nabla}\, \frac{1}{R_o} + \ldots \qquad (1\text{-}35)$$

where the neglected terms are of the second order or greater in the small quantity $(R_i/R_o)^2$. The sum

$$\vec{d} = \Sigma\, e_i\, \vec{R}_i \qquad (1\text{-}36)$$

is called the electric dipole moment of the system of charges. Note that if $\sum_i e_i = 0$, the dipole moment is independent of the choice of the origin of the coordinates. In this case the potential of the field at large distances is

$$\varphi = -\, \vec{d} \cdot \vec{\nabla}\, \frac{1}{R_o} \qquad (1\text{-}37)$$

Thus the potential of the field at large distances produced by a system of charges with total charge equal to zero is inversely proportional to the square of the distance and the field intensity αR_o^{-3}. This field has axial symmetry around the direction of $\vec{d}$.

The third term in the expansion of the potential in powers of $1/R_o$ is

$$\varphi^{(2)} = \frac{D_{\alpha\beta}\, n_\alpha\, n_\beta}{2\, R_o^3} \tag{1-38}$$

where

$$D_{\alpha\beta} = \sum_i (3\, x_{\alpha i}\, x_{\beta i} - r_i^2\, \delta_{\alpha\beta}) e_i \tag{1-39}$$

is the electric quadrupole moment of the system and n_α are the components of a unit vector along $\vec{R}_o$.

In a similar fashion we could write the succeeding terms of the expansion of φ, using the theory of spherical harmonics. For the

At a distance which is large compared with the dimensions of the system, the vector potential of the fields produced by all the charges at the point having the radius vector $\vec{R}_o$ is

$$\vec{A} = (\vec{M} \times \vec{R}_o) / R_o^3 \tag{1-40}$$

where

$$\vec{M} = \frac{1}{2c} \Sigma_i\, e_i\, \vec{R}_i \times \vec{v}_i \tag{1-41}$$

is the magnetic dipole moment of the system.

1.4 The Field of a Uniformly Moving Charge

Consider a charge e moving uniformly with velocity $\vec{v}$ along the x-axis in the laboratory frame of reference K. The charge is at rest in the frame K' which is moving with a velocity $\vec{v}$ along the x-axis of K. The axes y and z are parallel to y' and z'. At time t = 0 the origins of the two systems coincide and the charge is at its closest distance to the observer, who is located at the point P which has the coordinates (0,b,0) in the K frame. In the frame K' the observer's point P has the coordinates (- vt, b',0) and is a distance $R' = (b^2 + (vt')^2)^{\frac{1}{2}}$ away from the charge. In the rest frame K' the electric and magnetic fields are

$$\vec{E}' = e\vec{R}'/\vec{R}'^3 \qquad \vec{B}' = 0 \tag{1-42}$$

The coordinates in the two reference frames are related by the <u>Lorentz transformation</u>

$$x' = \gamma(x-vt);\ y' = y;\ z' = z;\ t' = \gamma\left(t- \frac{\beta x}{c}\right) \tag{1-43}$$

where

$$\beta = v/c\ ; \qquad \gamma = (1 - \beta^2)^{-\frac{1}{2}} \tag{1-44}$$

In terms of the coordinates of K the components of the electric field in K' is given by

$$E_x' = \frac{-\,e\gamma vt}{(b^2 + \gamma^2 v^2 t^2)^{3/2}}\ ,\ E_y' \frac{eb}{(b^2 + \gamma^2 v^2 t^2)^{3/2}} \tag{1-45}$$

The components of the electric and magnetic field parallel and perpendicular to the direction of motion of K' relative to K are given by the Lorentz transformation for the fields:

$$\vec{E}_{\parallel} = \vec{E}'_{\parallel} \qquad \vec{B}_{\parallel} = \vec{B}'_{\parallel}$$

$$\vec{E}_{\perp} = \gamma(\vec{E}'_{\perp} - \vec{\beta} \times \vec{B}') \qquad \vec{B}_{\perp} = \gamma(\vec{B}'_{\perp} + \vec{\beta} \times \vec{E}') \tag{1-46}$$

In this case we have $B' = 0$ so

$$E_x = E_x' = \frac{-e\gamma vt}{(b^2 + \gamma^2 v^2 t^2)^{3/2}} \qquad B_x = 0$$

$$E_y = \gamma E_y' = \frac{\gamma eb}{(b^2 + \gamma^2 v^2 t^2)^{3/2}} \qquad B_y = 0 \tag{1-47}$$

$$E_z = 0 \qquad B_z = \gamma\beta E_y' = \beta E_y$$

Introducing the angle θ between the direction of motion and the radius vector $\vec{R}$ from the charge e to the field point (x,y,z), we can write the expression for $\vec{E}$ in another form

$$\vec{E} = \frac{e\vec{R}}{R^3} \frac{(1 - \beta^2)}{(1 - \beta^2 \sin^2\theta)^{3/2}} \tag{1-48}$$

Along the direction of motion ($\theta = 0,\pi$) the field has the smallest value, equal to

$$\vec{E}_{\parallel} = e(1 - \beta^2) / R^2 \tag{1-49}$$

The largest field is for $\theta = \pi/2$:

$$\vec{E}_{\perp} = \gamma\, e/R^2 \tag{1-50}$$

Note that as the velocity increases, the field $\vec{E}_{\parallel}$ decreases, while $\vec{E}_{\perp}$ increases. For velocities close to the velocity of light, the denominator in (1-48) is close to zero in a narrow interval of values θ around $\theta = \pi/2$, with a width of the order

$$\Delta\theta \sim \gamma^{-1} \tag{1-51}$$

so that the electric field of a relativistic charge is large only in a narrow range of angles in the neighborhood of the equatorial plane. Thus as γ increases the peak fields increase $\alpha\gamma$, but the duration of the peak field at the field point decreases $\alpha\gamma^{-1}$. For large γ the observer sees nearly equal transverse and mutually perpendicular electric and magnetic fields, which are indistinguishable from a pulse of plane polarized radiation propagating in the x direction.

1.5 The Wave Equation

In a vacuum $\rho = 0$, and $\vec{j} = 0$ so Maxwell's equations become

$$\text{curl } \vec{E} = -\frac{1}{c}\frac{\partial \vec{B}}{\partial t} \tag{1-52}$$

$$\text{curl } \vec{B} = \frac{1}{c}\frac{\partial \vec{E}}{\partial t} \tag{1-53}$$

$$\text{div } \vec{E} = 0 \tag{1-54}$$

$$\text{div } \vec{B} = 0 \tag{1-55}$$

These equations possess non-zero solutions, so an electromagnetic field can exist even in the absence of any charges. Such fields are called electromagnetic waves, since they must necessarily by time-varying. Otherwise the solution, given by (1-26) and (1-33) with $\rho = 0$, and $\vec{j} = 0$ is $\varphi = 0$, $\vec{A} = 0$.

In general the vector potential $\vec{A}$ and the scalar potential φ are defined by the equations (cf. equations (1-24) and (1-30) for constant fields)

$$\vec{B} = \text{curl } \vec{A} \quad , \qquad \vec{E} = -\frac{1}{c}\frac{\partial \vec{A}}{\partial t} - \vec{\nabla}\varphi \qquad (1\text{-}56)$$

Equations (1-53) and (1-54) then become

$$\frac{1}{c^2}\frac{\partial^2 \vec{A}}{\partial t^2} - \nabla^2 \vec{A} + \vec{\nabla}(\vec{\nabla} \circ \vec{A} + \frac{1}{c}\frac{\partial \varphi}{\partial t}) = 0 \qquad (1\text{-}57)$$

$$\nabla^2 \varphi + \frac{1}{c}\vec{\nabla} \cdot \frac{\partial \vec{A}}{\partial t} = 0 \qquad (1\text{-}58)$$

It is desirable to decouple these equations. One way to do this is to choose $\vec{A}$, φ, such that they satisfy the Lorentz condition

$$\text{div } \vec{A} + \frac{1}{c}\frac{\partial \varphi}{\partial t} = 0 \qquad (1\text{-}59)$$

In this case the equations for the potentials become

$$[\nabla^2 - \frac{1}{c^2}\frac{\partial^2}{\partial t^2}]\, \vec{A} = -\, 4\pi\vec{j}/c = 0 \qquad (1\text{-}60)$$

$$[\nabla^2 - \frac{1}{c^2}\frac{\partial^2}{\partial t^2}]\, \varphi = -\, 4\pi\rho = 0 \qquad (1\text{-}61)$$

It is always possible to find potentials

to satisfy the Lorentz condition, because the vector potential is arbitrary to the extent that the gradient of some scalar function X can be added. Thus $\vec{B}$ is left unchanged by the transformation

$$\vec{A} \rightarrow \vec{A}' = \vec{A} + \nabla X \tag{1-62}$$

since curl (grad X) = 0. In order that the electric field be left unchanged by this transformation we must simultaneously have

$$\varphi \rightarrow \varphi' = \varphi - \frac{1}{c}\frac{\partial X}{\partial t} \tag{1-63}$$

The demand that $\vec{A}$ and φ satisfy the Lorentz condition can now be satisfied by choosing X appropriately. For example, if div A + $(1/c)\ \partial\varphi/\partial t = Y$ for one choice of $\vec{A}$ and φ, the transformations (1-62) and (1-63) can be used to obtain new potentials satisfying the Lorentz condition, provided

$$(\nabla^2 - \frac{\partial^2}{\partial t^2})\ X = -\ Y \tag{1-64}$$

The different possible choices one can make for $\vec{A}$ and φ, leaving $\vec{B}$ and $\vec{E}$ unchanged are called <u>gauges</u>, and the transformations (1-62) and (1-63) are called <u>gauge transformations</u>. The invariance of $\vec{E}$ and $\vec{B}$ under these transformations is called <u>gauge invariance</u>. The class of gauges satisfying (1-59) is called the <u>Lorentz gauge</u>. Another important gauge is the <u>Coulomb</u> or <u>transverse gauge</u>. Here one chooses X such that

$$\text{div}\ \vec{A} = 0 \tag{1-65}$$

$$\varphi = 0 \qquad (1\text{-}66)$$

for free space. The name, Coulomb gauge, is due to the fact that the scalar potential φ is given by the instantaneous Coulomb potential due to the charge density ρ in the general case. It is also called the transverse gauge because the condition (1-65) ensures that the fields are always perpendicular to the direction of propagation.

For plane waves propagating along the + x axis the fields are functions only of $t - x/c$. Therefore if the plane wave is monochromatic, its fields are simply periodic functions of $t - x/c$:

$$f = C \cos \omega(t - \frac{x}{c}) + D \sin \omega(t - \frac{x}{c}) \qquad (1\text{-}67)$$

here ω is the frequency in radians/sec and f denotes the scalar potential φ or one of the components of the vector potential $\vec{A}$, or one of the components of the electric or magnetic fields. It is usually more convenient to write the fields as the real parts of complex expressions:

$$f = \mathrm{Re}\, \{ f_o \exp(- i\omega(t - \frac{x}{c})) \} \qquad (1\text{-}68)$$

where f_o is a constant complex number.

The period of variation of the field with the coordinate x at a fixed time t is called the wavelength and is here denoted by λ:

$$\lambda = 2\pi c/\omega \qquad (1\text{-}69)$$

The quantity

$$k = \omega/c \tag{1-70}$$

is called the <u>wave number</u>. For the general case of propagation in an arbitrary direction x is replaced by the radius vector $\vec{R}$ and the wave number is replaced by the <u>wave vector</u>:

$$\vec{k} = \omega\hat{n}/c \tag{1-71}$$

where $\hat{n}$ is the unit vector along the direction of propagation of the wave:

$$\hat{n} = \vec{R}/R \tag{1-72}$$

Rewriting (1-68) in terms of the wave vector, we have

$$f = \text{Re} \{ f_o \exp(i\vec{k}\cdot\vec{R} - i\omega t) \tag{1-73}$$

The quantity $\vec{k}\cdot\vec{R}$ is called the <u>phase</u> of the wave. As long as we perform only linear operations, we can omit the sign Re for taking the real part and operate with complex quantities. Thus the expression for the vector potential of a plane, monochromatic wave can be written simply as

$$\vec{A} = \vec{A}_o \exp \{ i (\vec{k} \cdot \vec{R} - \omega t) \} \tag{1-74}$$

Substituting into equation (1-56) we find

$$\vec{E} = i k\vec{A} \quad ; \qquad \vec{B} = i\vec{k} \times \vec{A} \tag{1-75}$$

i.e., the electric and magnetic fields in a monochromatic plane wave are perpendicular to each other and to the direction of propagation of the wave.

1.6 Doppler Effect

Introduce the four dimensional wave vector with components

$$k_\mu = (\vec{k}, i\omega/c) \tag{1-76}$$

The phase of a monochromatic wave is just the scalar product of k_μ with the four-vector $x_\mu = (\vec{R}, ict)$,

$$k_\mu x_\mu = \vec{k} \cdot \vec{R} - \omega t \tag{1-77}$$

According to special relativity such scalar products are invariant under Lorentz transformations. Therefore for two frames in relative motion along the x-axis with velocity v the phase of a wave is the same:

$$\vec{k}' \cdot \vec{R}' - \omega' t' = \vec{k} \circ \vec{R} - \omega t \tag{1-78}$$

Using a Lorentz transformation to express $\vec{R}'$ and t' in terms of $\vec{R}$ and t and equating coefficients of the components of $\vec{R}$ and t on both sides of the equation we find

$$k_y' = k_y \qquad k_z' = k_z$$

$$k_x' = \gamma(k_x - \frac{v}{c^2}\omega) \tag{1-79}$$

$$\omega' = \gamma(\omega - v\, k_x)$$

For light waves $|k| = \omega/c$, $|k'| = \omega'/c$, so

$$\omega' = \gamma\omega(1 - \beta \cos\theta) \tag{1-80}$$

where θ is the angle between the direction of $\vec{k}$ and $\vec{v}$. It is related to the angle θ'

between $\vec{k}'$ and $\vec{v}'$ by

$$\tan\theta' = \sin\theta/\gamma(\cos\theta - \beta) \tag{1-81}$$

1.7 Polarization

For a monochromatic plane wave the electric field $\vec{E}$ is given by

$$\vec{E} = \vec{E}_o\, e^{i(\vec{k}\cdot\vec{R} - \omega t)} \tag{1-82}$$

where $\vec{E}_o$ is a complex vector. Suppose

$$\vec{E}_o = \vec{E}_1 + i\vec{E}_2 = (E_{1y} + iE_{2y})\hat{y} + (E_{1z} + iE_{2z})\hat{z} \tag{1-83}$$

Then (if we suppress the spatial dependence for the moment)

$$\begin{aligned} E_y &= E_{1y}\cos\omega t + E_{2y}\sin\omega t \\ E_z &= E_{1z}\cos\omega t + E_{2z}\sin\omega t \end{aligned} \tag{1-84}$$

Define α_1, α_2 by means of

$$\sin\alpha_1 = \frac{E_{1y}}{\sqrt{E_{1y}^2 + E_{2y}^2}} \qquad \sin\alpha_2 = \frac{E_{1z}}{\sqrt{E_{1z}^2 + E_{2z}^2}} \tag{1-85}$$

The equations (1-84) become

$$\begin{aligned} E_y &= E_{oy}\sin(\omega t + \alpha_1) \\ E_z &= E_{oz}\sin(\omega t + \alpha_2) \end{aligned} \tag{1-86}$$

where

$$E_{oy} = \sqrt{E_{1y}^2 + E_{2y}^2} \qquad E_{oz} = \sqrt{E_{1z}^2 + E_{2z}^2} \tag{1-87}$$

When the phase difference $\alpha_1 - \alpha_2 = \pi$, or some integral multiple thereof, the components E_y and E_z vary in phase, and the vector $\vec{E}$ traces out a straight line in the $E_y - E_z$ plane as t varies. The wave is said to be linearly polarized in this case. When the phase difference $= \pi/2$, or some odd integral multiple thereof, and $E_{oy} = E_{oz}$, $\vec{E}$ traces out a circle as t varies. The wave is said to be circularly polarized. If $E_{oy} \neq E_{oz}$ the wave is elliptically polarized. The wave is said to have right hand polarization if $\vec{E}$ rotates clockwise as seen by the observer and vice versa for left hand polarization.

If the major axis of the ellipse described by E_y and E_z makes an angle χ with the E_y axis, then

$$E_\chi = E_o \cos\xi \sin \omega t$$

$$E_{\chi+\pi/2} = E_o \sin\xi \cos \omega t \tag{1-88}$$

where

$$E_o^2 = E_{oy}^2 + E_{oz}^2 \tag{1-89}$$

and $\tan\xi$ is the ratio of the axes of the ellipse. The angles χ and ξ are related to α_1 and α_2 as follows:

$$\tan\alpha_1 = -\tan\xi \tan\chi$$

$$\tan\alpha_2 = \tan\xi \cot\chi \tag{1-90}$$

The case $\xi = 0$ corresponds to linear polarization, the case $\xi = \pi/4$ to circular polarization. Still another group of parameters is useful in practice for analyzing the polarization of a wave. They are the Stokes parameters, defined by (see Chandrasekhar, 1960):

$$I = E_{oy}^2 + E_{oz}^2 = E_o^2$$

$$Q = E_{oy}^2 - E_{oz}^2 = E_o^2 \cos 2\xi \cos 2\chi$$

$$U = 2E_{oy}E_{oz}\cos(\alpha_2 - \alpha_1) = E_o^2 \cos 2\xi \sin 2\chi$$

$$V = 2E_{oy}E_{oz} \sin(\alpha_2 - \alpha_1) = E_o^2 \sin 2\xi \tag{1-91}$$

Linear polarization implies $U = V = 0$, whereas for circular polarization $Q = U = 0$.

In practice the amplitudes and phases are not constants; however, due to the high frequency of vibration, we may assume that the amplitudes and phases are constant for many vibrations and yet change irregularly many times during the period of observation. The Stokes parameters may then be defined as a time average over many vibrations:

$$I = \overline{E_{oy}^2 + E_{oz}^2} \quad ; \text{ etc.} \tag{1-92}$$

This has the consequence that, for a number of independent waves, the Stokes parameters for the mixture is the sum of the respective

Stokes parameters of the separate streams:

$$I = \sum_i I_i \text{ ; etc.} \tag{1-93}$$

For an arbitrarily polarized beam, there always exists among the quantities I, Q, U and V the inequality

$$I^2 \geq Q^2 + U^2 + V^2 \tag{1-94}$$

The equality holds for the case when the ratio of the amplitudes and the difference in phase remain constant through all fluctuations. These are the same as the conditions for the radiation to be elliptically polarized.

The degree of elliptical polarization Π is defined as the ratio

$$\Pi = (Q^2 + U^2 + V^2)^{1/2}/ I \tag{1-95}$$

For circular polarization, $\Pi = V/I$, whereas for linear polarization $\Pi = Q/I$.

1.8 The Lienard-Wiechert Potentials

In Section 1.5 we saw that in the general case of non-zero charge and current density the vector and scalar potentials satisfy the equations

$$\nabla^2 \vec{A} - \frac{1}{c^2} \frac{\partial^2 \vec{A}}{\partial t^2} = - 4\pi\vec{j}/c \tag{1-96}$$

$$\nabla^2 \varphi - \frac{1}{c^2} \frac{\partial^2 \varphi}{\partial t^2} = - 4\pi\rho \tag{1-97}$$

The solution of the inhomogeneous equations (1-96) and (1-97) is

$$f = f_c + f_p \tag{1-98}$$

where f represents any one of the components of $\vec{A}$ or φ, f_c is the solution of the equation with the right-hand side equal to zero (complementary solution) and f_p is a particular integral of the equation. To find the particular integral we introduce the Green's function, G, defined by

$$\nabla^2 G - \frac{1}{c^2}\frac{\partial^2 G}{\partial t^2} = -4\pi\delta(\vec{R}_o - \vec{R}')\delta(t - t') \tag{1-99}$$

Note: The δ-function δ(x) is defined so that $\delta(x) = 0$ for all $x \neq 0$, and $\delta(x) \to \infty$ as $x \to 0$, so that the integral over a finite interval including $x = 0$ is equal to unity:

$$\int_{-b}^{b} \delta(x)\, dx = 1$$

where b is any non-zero number.
Therefore for any continuous function f(x),

$$\int_{-b}^{b} f(x)\delta(x)\, dx = f(0)$$

and

$$\int g(x)\, \delta\{f(x)\}\, dx = \{g(x)/f'(x)\}_{f(x) = 0}$$

where $f'(x) = df(x)/dx$.
Another useful equality is

$$\delta(\omega) = (1/2\pi)\int_{-\infty}^{\infty} e^{i\omega t}\, dt$$

The three-dimensional δ-function $\delta(\vec{R})$ is defined as

$$\delta(\vec{R}) = \delta(x)\delta(y)\delta(z)$$

Physically $G(\vec{R}_o, t, \vec{R}', t')$ represents the disturbance at $\vec{R}_o$ caused by a point source at $\vec{R}'$ turned on for only an infinitesimal interval at $t' = t$. Because of the linearity of the field equations, the actual field will be the sum of the fields produced by all such point sources (see Figure 1.1).

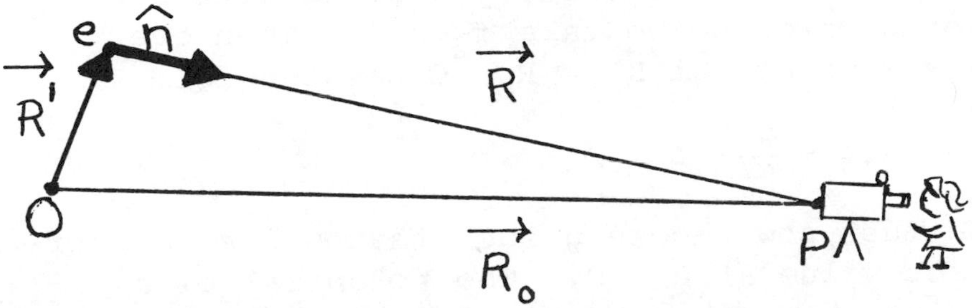

Figure 1.1.

The particular solution for φ, for example, is then given by

$$\varphi_\rho = \int d^3 R' \int dt\ G(\vec{R}_o, t, \vec{R}', t')\rho(\vec{R}', t) \qquad (1\text{-}100)$$

Everywhere except at $\vec{R}_o = \vec{R}'$, $\delta(\vec{R}_o - \vec{R}') = 0$ so we have

$$\nabla^2 G - \frac{1}{c^2}\frac{\partial^2 G}{\partial t^2} = 0 \tag{1-101}$$

Since in this case G should be spherically symmetric about $\vec{R}'$, we can assume that it is a function only of $R = |\vec{R}_o - \vec{R}'|$. Upon making the substitution $G = g/R$ we find

$$\frac{\partial^2 g}{\partial R^2} - \frac{1}{c^2}\frac{\partial^2 g}{\partial t^2} = 0 \tag{1-102}$$

The solution to this equation has the form

$$g = f_1\left(t - \frac{R}{c} - t'\right) + f_2\left(t + \frac{R}{c} - t'\right) \tag{1-103}$$

Since we only want a particular solution we need use only one of f_1 and f_2. In this situation we are clearly dealing with outgoing waves so we take $f_2 = 0$. Then everywhere except at $\vec{R}' = \vec{R}_o$, G has the form

$$G = g\left(t - \frac{R}{c} - t'\right) / R\ . \tag{1-104}$$

We must now choose g such that G has the correct value at R = 0. The potential of a point charge becomes infinite at the locus of the charge and so do its spatial derivatives, so the time derivatives in G can be neglected at $\vec{R}' = \vec{R}_o$, and (1-99) becomes

$$\nabla^2 G = -4\pi\delta(R)\delta(t - t') \tag{1-105}$$

This is just the Poisson equation (1-25) with φ replaced by G and ρ replaced by $\delta(R)\delta(t-t')$. The solution at the origin is therefore (see equation (1-27)) $G = \delta(t - t')/R$ and

$g = \delta(t - \frac{R}{c} - t')$. The Green's function for this problem is

$$G = \delta(t - \frac{R}{c} - t')/R \tag{1-106}$$

The general solutions of equations (1-96) and (1-97) are

$$\varphi = \int \frac{\delta(t - \frac{R}{c} - t')\rho(\vec{R}', t')}{R} dt' \; dV' + \varphi_o \tag{1-107}$$

$$\vec{A} = \int \frac{\delta(t - \frac{R}{c} - t')\vec{j}(\vec{R}', t')}{R} dt' \; dV' + \vec{A}_o \tag{1-108}$$

where φ_o and $\vec{A}_o$ are solutions of the homogeneous wave equation. These solutions are determined by the initial or boundary conditions. $R = |\vec{R}_o - \vec{R}'|$ is the distance between the source coordinate $\vec{R}'$ and the point $\vec{R}_o$ at which the field is observed. The potentials are sometimes called the retarded potentials because they exhibit the causal behavior associated with a wave disturbance. The effect observed at the point $\vec{R}_o$ at time t is due to the behavior of the current or charge density at an earlier or retarded time $t' = t - R/c$ at the point $\vec{R}'$.

For a point charge with a position vector $\vec{r}(t')$ and velocity vector $\vec{v}(t')$ the charge and current densities are given by

$$\rho(\vec{R}', t') = e \; \delta\{\vec{R}' - \vec{r}(t')\} \tag{1-109}$$

$$\vec{j}(\vec{R}',t') = e\,\vec{v}\,\delta\{\vec{R}' - \vec{r}(t')\} \tag{1-110}$$

Upon substituting these expressions into equations (1-107) and (1-108) and performing the integrations, using the properties of the δ-function given in the note following equation (1-99) the potentials are found to be (see Jackson, 1962):

$$\varphi = \{e/\kappa R\}_{ret} \tag{1-111}$$

$$\vec{A} = \{e\vec{v}/\kappa R\}_{ret} \tag{1-112}$$

where $\{\ \}_{ret}$ means that the quantity inside the brackets is to be evaluated at the retarded time $t' = t - R/c$, and

$$\kappa = 1 - \hat{n}\cdot\vec{\beta} \tag{1-113}$$

$$\hat{n} = \vec{R}/R \tag{1-114}$$

For non-relativistic motion $\kappa \to 1$. For relativistic motion κ becomes small for some angles, which implies large potentials.

By expanding the charge and current density into monochromatic waves, the potentials can be expressed in terms of Fourier components:

$$\varphi(\omega) = \frac{e}{2\pi}\int_{-\infty}^{\infty} \frac{e^{i\omega(t+R/c)}}{R}\,dt \tag{1-115}$$

and

$$\vec{A}(\omega) = \frac{e}{2\pi}\int_{-\infty}^{\infty} \frac{\vec{\beta}e^{i\omega(t+R/c)}}{R}\,dt \tag{1-116}$$

where

$$\varphi(t) = \int_{-\infty}^{\infty} \varphi(\omega)e^{-i\omega t}\, d\omega, \text{ etc.} \tag{1-117}$$

In the case of periodic motion an expansion in Fourier series yields:

$$\varphi(s) = \frac{2e}{\tau}\int_0^{\tau} \frac{e}{R}^{is\omega_0(t+R/c)}\, dt \tag{1-118}$$

and

$$\vec{A}(s) = \frac{2e}{\tau}\int_0^{\tau} \frac{\vec{\beta}e}{R}^{is\omega_0(t+R/c)}\, dt \tag{1-119}$$

where

$$\varphi(t) = \text{Re}\left\{\sum_{s=1}^{\infty} \varphi(s)e^{-is\omega_0 t}\right\} \tag{1-120}$$

and τ is the fundamental period of the motion ($= 2\pi/\omega_0$).

To determine the electric and magnetic fields, we need to differentiate the potentials with respect to position and time (see equation (1-56)). To do this it is simpler to work with the integral expressions for the potentials (1-107) and (1-108). Using the relationships

$$\frac{d\hat{n}}{dt'} = c\left[\frac{\hat{n} \times (\hat{n} \times \beta)}{R}\right] \tag{1-121}$$

and

$$\frac{1}{c}\frac{d}{dt'}(\kappa R) = \beta^2 - \vec{\beta}\circ\hat{n} - \frac{R}{c}\,\hat{n}\cdot\dot{\vec{\beta}} \qquad (1\text{-}122)$$

one finds

$$\vec{E}(\vec{R}_o, t) = e\left[\frac{(\hat{n}-\vec{\beta})(1-\beta^2)}{\kappa^3 R^2}\right]_{ret} + \frac{e}{c}\left[\frac{\hat{n}}{\kappa^3 R}\times\{(\hat{n}-\vec{\beta})\times\dot{\vec{\beta}}\}\right]_{ret}$$

$$\vec{B} = \hat{n}\times\vec{E} \qquad (1\text{-}123)$$

The fields consist of two types. The first depends only on the velocity of the particle and not its acceleration and varies at large distances as $1/R^2$. The second depends on the acceleration and varies as $1/R$ at large distances. The ratio of the two types of fields is

$$|E_{vel}/E_{acc}| \sim \frac{c}{R_o}\frac{(1-\beta^2)}{\dot{\beta}} \sim \frac{c\tau}{R_o}(1-\beta^2)$$

$$\sim \frac{\lambda}{R_o}(1-\beta^2) \qquad (1\text{-}124)$$

where τ is the characteristic time for changes in the system and λ is the characteristic wavelength of radiation from the system. Thus at distances R_o large compared to the wavelength of the radiation, $E_{vel}/E_{acc} \ll 1$, the electric and magnetic fields are given by

$$\vec{E}(\vec{R}, t) = \frac{e}{c}\left[\frac{\hat{n}\times\{(n-\vec{\beta})\times\dot{\vec{\beta}}\}}{\kappa^3 R}\right]_{ret}$$

$$\vec{B}(\vec{R},t) = \hat{n} \times \vec{E} \tag{1-125}$$

Since the point of observation is assumed to be at a large distance compared with the region where the charge changes its direction, ($R_o >> R'$), the vector $\vec{R}$ is approximately in the same direction as $\vec{R}_o$, and

$$R \simeq R_o \left(1 - \frac{\hat{n}\cdot\vec{R}'}{R_o}\right) \tag{1-126}$$

To determine the potentials from equations (1-111) and (1-112) or the fields from equation (1-125), we can neglect $\hat{n}\circ\vec{R}'/R_o$ compared with unity in the denominator, but not in evaluating the expressions at the retarded time. Whether or not this term can be neglected in that calculation depends not on the relative values of $\vec{R}_o$ and $\hat{n}\cdot\vec{R}'$, but on how much the velocity and acceleration changes in the time $\hat{n}\cdot\vec{R}'/c$.

Thus when the point of observation is at a large distance compared with the region where the charge carries out its motion, the potentials can be written in the form

$$\varphi = \frac{e}{R_o}\left[\frac{1}{1 - \hat{n}\cdot\vec{\beta}(t')}\right]_{t'=t - \frac{R_o}{c} + \frac{\hat{n}\circ\vec{R}'}{c}} \tag{1-127}$$

and

$$\vec{A} = \frac{e}{R_o}\left[\frac{\vec{\beta}}{1 - \hat{n}\cdot\vec{\beta}(t')}\right]_{t' = t - \frac{R_o}{c} + \frac{\hat{n}\circ\vec{R}'}{c}} \tag{1-128}$$

The Fourier components become (see equations (1-115), (1-116))

$$\varphi(\omega) = \left(\frac{e}{2\pi R_o}\right) e^{ikR_o} \int e^{i[\omega t - \vec{k}\circ\vec{R}']} dt \qquad (1\text{-}129)$$

and

$$\vec{A}(\omega) = \left(\frac{e}{2\pi R_o}\right) e^{ikR_o} \int \vec{\beta}\, e^{i[\omega t - \vec{k}\circ\vec{R}']} dt \qquad (1\text{-}130)$$

1.9 Dipole Radiation

The term $\hat{n}\circ\vec{R}'/c$ in the retarded time can be neglected if the distribution of charge changes by a negligible amount during that time.

If r is the characteristic dimension of the system then

$$\hat{n}\circ\vec{R}'/c \sim r/c \qquad (1\text{-}131)$$

If the time scale for an appreciable change in position of the charge is τ, then the term $\hat{n}\cdot\vec{R}'/c$ will be small if

$$r \ll c\tau \qquad (1\text{-}132)$$

But τ is related to the frequency of the radiation from the system by $\tau \sim 1/\nu$, so the condition (1-132) can also be written as

$$r \ll c/\nu = \lambda \qquad (1\text{-}133)$$

That is, the dimensions of the system must be small compared to the wavelength of the radiation. From the definition of (1-70) of the wave number, this is also equivalent to

$kR' \ll 1$ (1-134)

The condition for the neglect of the term $\hat{n}\cdot\vec{R}'/c$ can also be expressed in terms of the velocity $\vec{v}$ of the charges, which must be of order r/τ, from which it follows, using (1-133) that

$v \ll c$ (1-135)

i.e., the motion should be non-relativistic.

This approximation is called the dipole approximation and the radiation in this case is called dipole radiation. In this approximation we can set $\kappa = 1$, $R = R_o$ and evaluate all quantities at $t' = t - R_o/c$, which is a considerable simplification since R_o is independent of t'. For all practical purposes we can drop the reference to the retarded time when working in the dipole approximation.

The expression for the electric field then takes the form

$$\vec{E} = \frac{e}{c}\,\hat{n} \times (\hat{n} \times \dot{\vec{\beta}})/cR \tag{1-136}$$

For a number of charges

$$\vec{E} = \hat{n} \times \frac{(\hat{n} \times \Sigma\, e\dot{\vec{\beta}})}{cR} = \hat{n} \times \frac{(\hat{n} \times \ddot{\vec{d}})}{c^2 R} \tag{1-137}$$

where $\vec{d}$ is the dipole moment of the system (see equation (1-36)). Note that the radiation is determined by the second derivative of the dipole moment, hence the name "dipole approximation".

The power radiated is obtained from

Poynting's vector (see equation (1-15))

$$\vec{S} = c\,\frac{(\vec{E}\times\vec{B})}{4\pi} = (\ddot{d})^2 \sin^2\Theta\,\frac{\hat{n}}{4\pi c^3 R^2} \qquad (1\text{-}138)$$

where Θ is the angle between $\ddot{\vec{d}}$ and $\hat{n}$ and $d = |\vec{d}|$.

The power radiated per unit solid angle is given by

$$\frac{dP}{d\Omega} \equiv P(\Omega) = R^2\ \vec{S}\cdot\hat{n} = (\ddot{d})^2\,\frac{\sin^2\Theta}{4\pi c^3} \qquad (1\text{-}139)$$

Upon integration of equation (1-139) over solid angle $d\Omega = 2\pi\ \sin\Theta\ d\Theta$, the total power radiated is found to be

$$\frac{dW}{dt} \equiv P = \frac{2(\ddot{d})^2}{3c^3} \qquad (1\text{-}140)$$

This is <u>Larmor's formula</u> for the radiation from a non-relativistic charge. Note that the angular distribution of the radiation is symmetric about the direction of the acceleration of the charge and independent of its velocity (see equation (1-139) and Figure 1.2).

To obtain information about the spectrum of dipole radiation, we need the Fourier components of $dP/d\Omega$. We cannot calculate this, but we can calculate the energy $dW/d\Omega$ radiated per unit solid angle over the entire time during which the charge is accelerated:

$$\frac{dW}{d\Omega} = \int (dP/d\Omega)\,dt = (cR^2/4\pi)\int E^2\,dt \qquad (1\text{-}141)$$

by using Parseval's theorem:

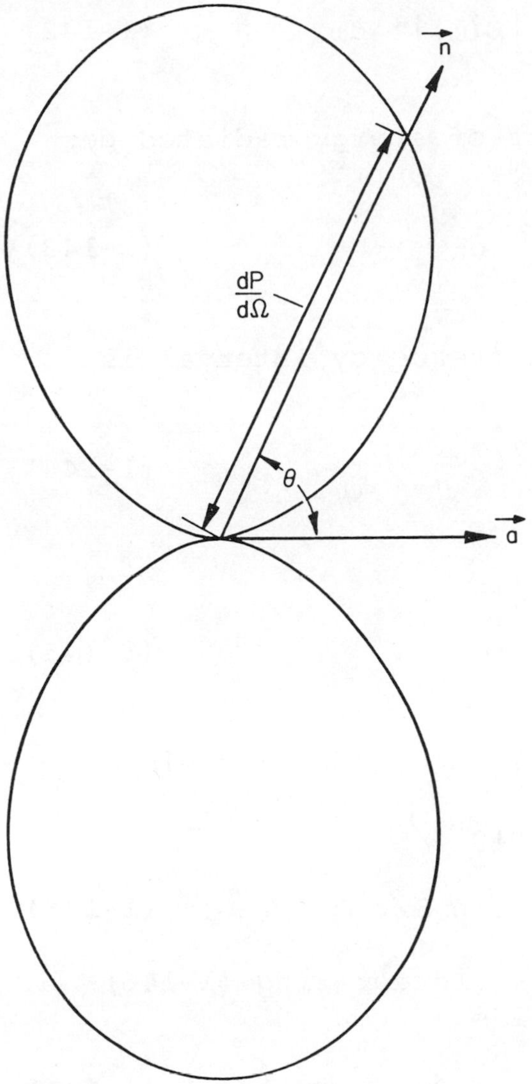

Figure 1.2. The angular distribution of the power radiated by a non-relativistic charge undergoing an acceleration a. The power unit solid angle radiated in the direction of the vector $\vec{n}$ is proportional to the radius vector as indicated in the figure.

$$\int_{-\infty}^{\infty} E^2(t)dt = 4\pi \int_0^{\infty} |E(\omega)|^2 \, d\omega \qquad (1\text{-}142)$$

Therefore the amount of energy radiated per solid angle is

$$\frac{dW}{d\Omega} = cR^2 \int_0^{\infty} |E(\omega)|^2 \, d\omega \qquad (1\text{-}143)$$

The energy per unit frequency interval is therefore

$$\frac{dW(\omega)}{d\Omega} = cR^2 \, |E(\omega)|^2 \qquad (1\text{-}144)$$

where

$$E(\omega) = \hat{n} \times \frac{(\hat{n} \times \ddot{\vec{d}}(\omega))}{c^2 R} \qquad (1\text{-}145)$$

Substituting (1-145) into (1-144) yields

$$dW(\omega)/d\Omega = [\ddot{d}(\omega)]^2 \sin^2\Theta/c^3$$

$$= \omega^4 \, d^2(\omega) \sin^2\Theta/c^3 \qquad (1\text{-}146)$$

since $\ddot{d}(\omega) = \omega^2 d(\omega)$. Integrating (1-146) over all angles yields

$$W(\omega) = 8\pi(\ddot{d}(\omega))^2/3c^3 = 8\pi \, \omega^4 d^2(\omega)/3c^3 \qquad (1\text{-}147)$$

For periodic motion Parseval's theorem takes the form

$$\frac{2}{T}\int_0^{T} E^2(t)dt = \sum_{s=1}^{\infty} |E_s|^2 \qquad (1\text{-}148)$$

so

$$\tau^{-1}(dW_s/d\Omega) = (\ddot{d}_s)^2 \sin^2\Theta/8\pi c^3 \qquad (1\text{-}149)$$

The classical formula for dipole radiation can be used to calculate low frequency radiation resulting from the acceleration of a charged particle (see Chapters 4 and 5).

1.10 Radiation from a Relativistic Charged Particle

In the non-relativistic case we derived a simple expression for the total power radiated by an accelerated charge (see equation (1-140))

$$\frac{dW}{dt} = \frac{2}{3}\frac{e^2 \dot{v}^2}{c^3} \qquad (1\text{-}150)$$

Of course this formula does not apply when the particle motion is relativistic. However in the frame of reference where the particle is at rest we certainly have $v << c$ and (1-150) applies. In this reference frame the particle radiates in time dt the energy

$$dW = \left(\frac{2e^2 \dot{v}^2}{3c^3}\right) dt \qquad (1\text{-}151)$$

In this reference frame the momentum radiated is zero:

$$d\vec{p} = \int \overleftrightarrow{T} \cdot \hat{n}\, dA\, dt = 0 \qquad (1\text{-}152)$$

(see (1-19) and (1-136)). This is due to the symmetry of dipole radiation.

Writing equations (1-151) and (1-152) in four-vector notation we have

$$dp_i = (2e^2/3c)\ (du_k/ds)^2\ dx_i \qquad (1\text{-}153)$$

where

$$u_i = dx_i/ds = \gamma v_i/c \qquad (1\text{-}154)$$

$$ds = c\ dt/\gamma \qquad (1\text{-}155)$$

$$v_i = (\vec{v},\ ic) \qquad (1\text{-}156)$$

Since dp_i and dx_i are both four-vectors, the quantity relating them must be a scalar and therefore a Lorentz invariant:

$$(2\ e^2/3c)\ (du_k/ds)^2 = \text{invariant} \qquad (1\text{-}157)$$

The total power radiated in an arbitrary reference frame is found by noting that

$$dp_4 = i\ dW/c \ ; \ dx_4 = i\ c\ dt \qquad (1\text{-}158)$$

so

$$\begin{aligned} dp_4/dx_4 &= (1/c^2)\ dW/dt \\ &= (2\ e^2/3c^3)\gamma^2\ (du_i/dt)^2 \qquad (1\text{-}159) \end{aligned}$$

Since this quantity is a Lorentz invariant we have that the total power radiated in any frame of reference (arbitrary velocity) is given by

$$dW/dt = (2\ e^2/3c)\gamma^2\ (du_i/dt)^2$$

$$= (2\ e^2/3c)\gamma^6\ \{\mathring{\beta}^2 - (\vec{\beta} \times \vec{\mathring{\beta}})^2\} \qquad (1\text{-}160)$$

In terms of forces, note that for a particle of mass m:

$$d\vec{u}/ds = (\gamma/mc^2)d\vec{p}/dt \qquad (1\text{-}161)$$

$$du_4/ds = (\gamma/mc^2)\ dW/dt = \gamma\ d\gamma/dt \qquad (1\text{-}162)$$

In the case of a particle in electric and magnetic fields

$$d\vec{u}/ds = (e\gamma/mc^2)\ (\vec{E} + \vec{\beta} \times \vec{B}) \qquad (1\text{-}163)$$

$$du_4/ds = e\gamma/mc^2)\ (\vec{\beta} \cdot \vec{E}) \qquad (1\text{-}164)$$

(see equations (1-7) and (1-8)).

$$dW/dt = (2\ r_o^2\ c/3)\gamma^2\ \{|\vec{E} + \vec{\beta}\times\vec{B}|^2 - |\vec{\beta}\cdot\vec{E}|^2\} \qquad (1\text{-}165)$$

where E, B refer to <u>external</u> fields, and

$$r_o = e^2/m\ c^2 \quad . \qquad (1\text{-}166)$$

Equation (1-165) shows that the power radiated is inversely proportional to the square of the mass of the radiating particle, so that electrons radiate much more energy than protons in given electric and magnetic fields.

In the case of a particle moving parallel to the magnetic field and experiencing acceleration by an electric field which is parallel to the magnetic field,

$$dW/dt = (2\ r_o^2 c/3)E^2$$

$$= (2\, r_o^2 c/3e^2)(d\vec{p}/dt)^2 \qquad (1\text{-}167)$$

In the case of motion in a magnetic field, with the electric field equal to zero,

$$dW/dt = (2\, r_o^2 c/3)\gamma^2\ \beta_\perp^2\ B^2$$

$$= (2\, r_o^2 c/3e^2)\gamma^2 (d\vec{p}/dt)^2 \qquad (1\text{-}168)$$

where $\beta_\perp$ is the component of $\vec{\beta}$ perpendicular to the magnetic field. For relativistic particles these losses are proportional to the square of the energy and can become very large.

Thus for comparable forces, the power radiated by relativistic charges is a factor γ^2 less for acceleration parallel to the velocity than for acceleration perpendicular to the velocity.

In order to determine the angular distribution of the radiation, we must substitute the fields given by equation (1-125) into equation (1-15):

$$dP(t)/d\Omega = cE^2 R^2/4\pi$$

$$= (e^2/4\pi c)\ \{|\hat{n}\times[(\hat{n}-\vec{\beta})\times\dot{\vec{\beta}}]|^2/\kappa^6\}_{t'+\frac{R}{c}=t} \qquad (1\text{-}169)$$

This is the energy per unit solid angle per unit time detected at the observer at time t due to radiation emitted by the charge at a time $t' = t - R/c$. To get the power radiated per unit solid angle in terms of the charge's time, we must multiply equation (1-169) by the factor (dt/dt') to take into account the relativistic effects caused by the charge's motion toward or away from the observer. From equation (1-126)

$$dt/dt' = 1 - \hat{n}\cdot\vec{\beta} = \kappa \tag{1-170}$$

In the ultrarelativistic case $\beta \simeq 1$, and $1 - \beta \ll 1$, so the terms in the denominator become small for $\hat{n}\cdot\vec{\beta} \simeq \beta$ i.e., for radiation in the direction of $\vec{\beta}$. If θ denotes the angle between $\hat{n}$ and $\vec{\beta}$ then for small θ

$$\kappa = 1 - \beta\cos\theta \approx 1 - \beta + \frac{\beta\theta^2}{2} \tag{1-171}$$

For $\beta \simeq 1$, the expansion on the right will be small if the third term is of the order of the first two, i.e., if

$$\theta^2 \simeq 2(1-\beta)/\beta \simeq (1+\beta)(1-\beta) = 1/\gamma^2$$

or

$$\theta \simeq 1/\gamma \tag{1-172}$$

Thus most of the radiation is confined to a narrow cone of half-angle $\simeq \gamma^{-1}$ around the direction of the velocity of the particle.

When the velocity and acceleration of the particle are parallel, the intensity distribution is

$$dP(t')/d\Omega = (e^2/4\pi c)\dot{\beta}^2 \sin^2\theta/(1-\beta\cos\theta)^5 \tag{1-173}$$

For $\beta \ll 1$, this reduces to the Larmor result. As the speed of the charge approaches the speed of light

$$dP(t')/d\Omega \underset{\beta\to 1}{\to} (8e^2/\pi c)\gamma^8\ \dot{\beta}^2\ (\gamma\theta)^2/(1+\gamma^2\theta^2)^5$$

(1-174)

As the velocity of the charge increases, the "figure eight" distribution characteristic of radiation from a non-relativistic charge is tipped forward and the peak intensity increases in magnitude proportional γ^8.

Integrating equation (1-174) over all angles, we obtain the result given in equation (1-160) for $\vec{\beta} \times \dot{\vec{\beta}} = 0$.

When the velocity and the acceleration are perpendicular

$$dP(t')/d\Omega = (e^2/4\pi c) \cdot \dot{\beta}^2 \left\{\frac{\gamma^2(1-\beta\cos\theta)^2 - \sin^2\theta\cos^2\varphi}{\gamma^2(1-\beta\cos\theta)^5}\right\}$$

(1-175)

where φ is the azimuthal angle of $\hat{n}$ relative to the plane passing through $\vec{\beta}$ and $\dot{\vec{\beta}}$. Again this reduces to the Larmor result for small β, since $1 - \sin^2\theta\cos^2\varphi = \sin^2\theta$. In the ultrarelativistic case (1-175) becomes

$$dP(t')/d\Omega = (2e^2/\pi c)\dot{\beta}^2 \cdot \gamma^6 \left\{\frac{(1+\gamma^2\theta^2)^2 - 4\gamma^2\theta^2\cos^2\varphi}{(1+\gamma^2\theta^2)^5}\right\}$$

(1-176)

The radiation pattern for the case $\varphi = 0$ and acceleration produced by the magnetic field (see Chapter 3) is shown below in Figure 1.3.

Figures 1.2 and 1.3, or equations (1-139), (1-173) and (1-176) illustrate how the radiation pattern from an accelerated charge changes as its velocity increases. For non-relativistic motion the angular distribution is independent of the velocity vector and is distributed over a wide angle. For relativistic motion the radiation is greatly enhanced in the direction of motion and is confined to a very narrow cone about that direction.

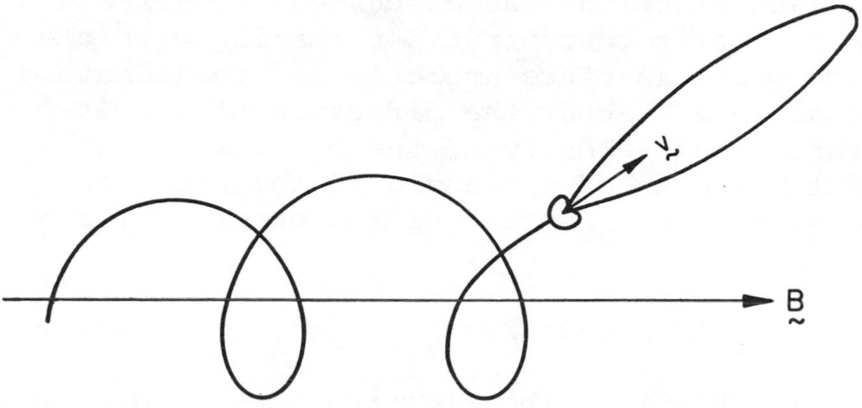

Figure 1.3. A relativistic particle spiraling in a magnetic field emitting synchrotron radiation with the angular pattern as indicated.

For a charged particle undergoing arbitrary ultra-relativistic motion the radiation emitted at any instant can be thought of as a coherent superposition of contributions coming from the components of acceleration parallel and perpendicular to the velocity. However, equations (1-167) and (1-168) show that, for comparable parallel and perpendicular forces the radiation from the parallel component is negligible (of order $1/\gamma^2$) compared to that from the perpendicular component. Therefore the radiation emitted by ultrarelativistic particles is very nearly the same as that emitted by a particle moving instantaneously along the arc of a circular path whose radius of curvature is given by

$$r_c = c^2/\dot{v}_\perp \tag{1-177}$$

where $\dot{v}_\perp$ is the perpendicular component of the acceleration. As discussed above the radiation is concentrated primarily within a cone whose aperture angle $\Delta\theta$ is approximately equal to $2/\gamma$ about the direction of the instantaneous velocity of the particle. Within the limits of the angle $\Delta\theta$ the electron moves in the direction of the observer for a time

$$\Delta t' \simeq r_c\, \Delta\theta/c \simeq 2r_c/c\gamma \tag{1-178}$$

During this time the electron moves a distance $v\Delta t'$ in the direction of the observer, so the radiation pulse contracts the length $v\Delta t'$. As a result, the observed length of the pulse is of the order

$$c\Delta t = (c-v)\Delta t' \tag{1-179}$$

and its duration is

$$\Delta t = \Delta t'(1-\beta) \simeq \Delta t'/\gamma^2 \tag{1-180}$$

The observed radiation spectrum will

therefore contain frequencies up to a maximum frequency ω_m:

$$\omega_m \simeq 1/\Delta t \simeq c\gamma^3/r_c \tag{1-181}$$

For frequencies much greater than this the exponential term in the Fourier transform of the fields oscillates rapidly and the slowly varying parts of the integral interfere destructively so that the integral becomes negligibly small (see equation (1-183)).

In the case of circular motion in a magnetic field

$$r_c = m_e c^2 \beta\, \gamma/eB \simeq c\gamma/\omega_B \tag{1-182}$$

where $\omega_B = eB/m_e c$ is the electron cyclotron frequency. The observed radiation spectrum will consist of harmonics of the frequency ω_B/γ extending up to $\omega_m \simeq \gamma^2 \omega_B$.

The equation describing the spectral distribution of the radiation from an accelerated charge can be obtained from equations (1-115), (1-125), (1-126), (1-144), and (1-170):

$$\frac{dW(\omega)}{d\Omega} = \frac{e^2}{4\pi^2 c}\left|\int_{-\infty}^{\infty} \frac{\hat{n} \times [(\hat{n}-\vec{\beta}) \times \dot{\vec{\beta}}]}{\kappa^2} \cdot e^{i\omega(t' - \hat{n}\circ R'/c)}\, dt'\right|^2 \tag{1-183}$$

This expression can be integrated by parts using the relationship

$$\frac{\hat{n} \times (\hat{n}-\vec{\beta}) \times \dot{\vec{\beta}}}{(1 - \hat{n}\cdot\vec{\beta})^2} = \frac{d}{dt'}\left\{\frac{\hat{n} \times (\hat{n}\times\vec{\beta})}{1 - \hat{n}\cdot\vec{\beta}}\right\}$$

The result is

$$dW(\omega)/d\Omega = (e^2\omega^2/4\pi^2 c) \cdot \left| \int_{-\infty}^{\infty} \hat{n} \times (\hat{n}\times\vec{\beta}) e^{i\omega(t' - \hat{n}\cdot\vec{R}'/c)} dt' \right|^2 \qquad (1\text{-}183')$$

In the case of periodic motion

$$dW_s/\tau d\Omega = dP|d\Omega = (e^2\omega^2/8\pi^3 c) \cdot \left| \int_0^{\tau} \hat{n} \times (\hat{n}\times\vec{\beta}) e^{is\omega_0(t-\hat{n}\cdot\vec{R}'/c)} dt' \right|^2 \qquad (1\text{-}183'')$$

1.11 The Influence of Cosmic Plasma on the Propagation and Emission of Electromagnetic Waves

Up to this point it has been assumed that the radiation is emitted and propagated in a vacuum. Usually, this is a reasonable approximation to the actual situation. Sometimes, however, the medium radically influences the character of the electromagnetic radiation, with regard to both the emission and propagation of the waves.

Radiation processes in a dielectric medium can be discussed in terms of the formalism developed for radiation in a vacuum by making the substitutions

$$c \rightarrow c/n_r \qquad\qquad e \rightarrow e/n_r \qquad (1\text{-}184)$$

where n_r is the index of refraction, which

for an isotropic plasma and for frequencies much greater than the frequency of collisions between particles is (Alfven and Falthammar, 1963, Ginzburg, 1964)

$$n_r^2(\omega) = 1 - (\omega_p^2/\omega^2) \qquad (1\text{-}185)$$

$$\omega_p = (4\pi\ N_e\ e^2/m_e)^{\frac{1}{2}} = (3\times 10^9\ N_e)^{\frac{1}{2}}\ \text{rad/sec} \qquad (1\text{-}186)$$

(N_e = electron number density).

That the transformation (1-184) is the correct one follows from the form of Maxwell's equations for a dielectric medium:

$$\vec{\nabla}\cdot\vec{D} = 4\pi\rho$$

$$\vec{\nabla}\times\vec{B} = 4\pi\vec{j}/c + \partial\vec{D}/c\partial t \qquad (1\text{-}187)$$

Since $\vec{D} = n_r^2\ \vec{E}$, it is evident that the substitution (1-184) will cast the equations into a form identical with the form of Maxwell's equations for a vacuum.

The angular and frequency distribution of radiation emitted by a charged particle in motion is given by

$$dW(\omega)/d\Omega = (e^2\,\omega^2\,n_r/4\pi^2 c) \cdot \left| \int_{-\infty}^{\infty} \hat{n}\times(\hat{n}\times\vec{\beta})\,e^{i\omega(t - n\ \hat{n}\cdot\vec{R}(t')/c)}\,dt \right|^2 \qquad (1\text{-}188)$$

When $\omega < \omega_p$, n_r is imaginary and the radiation is exponentially damped.

The index of refraction can also have an imaginary component if absorption is occurring in the plasma. Denoting the imaginary component due to absorption by μ:

$$n_r^* = n_r + i\,\mu \qquad (1\text{-}189)$$

The electric field produced by an accelerated charge is

$$\vec{E} = \vec{E}_o\, e^{-\mu\omega z/c}\; e^{i\omega(t-n_r z/c)} \qquad (1\text{-}190)$$

for a wave propagated in the z-direction. In general the index of refraction and the absorption coefficient depend on the properties of the medium (density, temperature, magnetic field) and the frequency of the radiation. The determination of the exact nature of the dependence is a complex problem to which entire books are devoted (see e.g. Ginzburg, 1964). The general method of computing the absorption coefficient from the properties of the plasma is discussed in Chapter 2.

If the plasma has a magnetic field B, and if absorption is unimportant, index of refraction takes the form

$$n^2_{o,x} =$$

$$1 - \frac{2V(1 - V)}{2(1-V)-U\sin^2\alpha \pm (U^2\sin^4\alpha + 4U(1-V)^2\cos^2\alpha)^{\frac{1}{2}}}$$

$$V = (\omega_p/\omega)^2 \qquad U = (\omega_B/\omega)^2 \qquad (1\text{-}191)$$

Here α is the angle between the direction of

propagation of the wave and the magnetic field. The subscripts o and x refer to the two possible modes of propagation, the ordinary and the extraordinary modes, corresponding to taking the positive and negative square root, respectively, in the denominator of (1-191).

When B = 0, equation (1-191) reduces to the simple form (1-185) and there is no distinction between the o- and x-modes. For $B \neq 0$ and propagation along the magnetic field ($\alpha = 0$), (1-191) becomes

$$n^2_{o,x} = 1 - (\omega_p^2/\omega(\omega \pm \omega_B)) \qquad (1\text{-}192)$$

Using (1-192) in the wave equation shows that the o- and x-modes correspond to circularly polarized waves rotating clockwise (o) and counterclockwise (x) as they propagate along the field.

For waves propagating perpendicular to the magnetic field ($\alpha = \pi/2$),

$$n^2_o = 1 - (\omega_p^2/\omega^2)$$

$$n^2_x = 1 - (\omega_p^2(\omega^2 - \omega_p^2)/\omega^2(\omega^2 - \omega_p^2 - \omega_B^2)) \qquad (1\text{-}193)$$

The o-mode is polarized parallel to the magnetic field. Thus the magnetic field has no effect on the motion of the charges and the velocity of propagation is independent of the strength of the magnetic field. The x-mode is polarized perpendicular to the field so its propagation velocity does depend on the field strength.

For most cases of interest the frequency of the wave is much greater than the electron cyclotron frequency ($\omega >> \omega_B$) and the expression (1-191) can be simplified. In the limit where

$$1 - (\omega_p^2/\omega^2) >> \omega_B \sin^2\alpha/\omega \cos\alpha \tag{1-194}$$

(1-191) reduces to

$$n_{o,x}^2 = 1 - (\omega_p^2/\omega(\omega \pm \omega_L)) \tag{1-195}$$

where $\omega_L = \omega_B \cos\alpha$. Equation (1-195) is the same as equation (1-192) for propagation along the magnetic field (longitudingal propagation) with ω_B replaced by ω_L. Hence the propagation is called "quasi-longitudinal". This approximation is adequate to describe most of the situations encountered in astrophysics.

The propagation of waves in a magnetoactive plasma generally depends strongly on the intensity and direction of the magnetic field. However, except in the vicinity of stars, the magnetic fields are sufficiently weak that the frequency of the radiation is much greater than the electron cyclotron frequency:

$$\omega >> \omega_B \tag{1-196}$$

In this limit the plasma can be considered practically isotropic, with the index of refraction given by (1-185). However, in the consideration of the polarization of the wave, even a small anisotropy can be important.

This is the basis of the Faraday effect, which is an important tool for measuring cosmic magnetic fields.

Consider a wave of amplitude A which is linearly polarized in the y-direction. This wave can be decomposed into two circularly polarized waves with opposite directions of rotation. At the origin we have

$$E = E_o + E_x = A\,e^{i\omega t} + Ae^{-i\omega t} \tag{1-197}$$

In the quasi-longitudinal approximation the two waves propagate at different velocities given by (1-195) so after propagation through a distance R the waves are described by

$$E_{o,x} = Ae^{\pm\, i\omega(t+n_{o,x}R/c)} \tag{1-198}$$

and the composite wave by

$$E = Ae^{i\Delta/2}\,(e^{i\omega(t+s)} + e^{-i\omega(t+s)}) \tag{1-199}$$

where

$$s = (n_o + n_x)(\omega R/c)$$

$$\Delta = (n_o - n_x)(\omega R/c) \tag{1-200}$$

Thus after traveling a distance R cm the wave is still linearly polarized but has been rotated through an angle

$$\begin{aligned}\psi = \Delta/2 &= \omega_p^2\omega_B R\cos\alpha/2\omega^2 c\\ &= 2.4\times10^4\ N_e B\ R\cos\alpha/\nu^2\ \ \text{rad.}\end{aligned} \tag{1-201}$$

$(\nu = \omega/2\pi)$.

In general, B, N_e and α will not be constant along the line of sight, so the product $N_e BR\cos\alpha$ must be replaced by the integral $\int N_e B\cos\alpha \, dR$, taken along the line of sight. If we express the frequency in terms of the wavelength of the radiation in meters, λ_m, and dR in parsecs, then

$$\psi = (8.1\times10^5 \int N_e B \cos\alpha \, dR)^2 \lambda_m^2 = R_m \lambda_m^2 \quad (1\text{-}202)$$

where R_m is called the rotation measure.

This integral cannot be determined from a single observation of the position angle of the plane of polarization because there is almost never any way to estimate the position angle of the plane of polarization at the source, and because there is no way to distinguish between values of ψ that differ by 180°. It is necessary to observe the source at several frequencies, and then to plot the observed position angles as a function of λ_m^2. The straight line fit to these points then gives the rotation measure.

Observations of the polarization of radio sources shows that the magnitudes of the Faraday rotation are on the average much smaller for high latitude extragalactic radio sources than for low latitude sources. Thus it seems that the major part of the rotation occurs within the galaxy rather than in the sources themselves or in the intergalactic medium. In addition it has been observed (Morris and Berge, 1964) that the sense of Faraday rotation changes sign from one side of the galactic plane to the other. This indicates that, in the neighborhood of the sun, the magnetic field changes sign when crossing the galactic plane. Magnetic fields calculated from the absolute value of the rotation measure range from 10^{-6} gauss to a few times 10^{-5} gauss, depending on the assumed values of N_e and R, the distribution of electron density over the field structures, and whether the field is predominantly uniform in direction, or is composed of a number of anti-parallel filaments, or is rather irregular.

The radiation from cosmic radio sources can be de-polarized if the rotation measure is not the same for all the elements of the source within the observing beamwidth. These effects are discussed by Gardner and Whiteoak (1966) and Burn (1966).

References

Good general references for the material in this chapter are:

Jackson (1962) Chapters 6, 14, 15.

Landau and Lifshitz (1962) Chapters 4, 5, 6, 8 and 9.

Other references for particular topics:

Polarization:

Chandrasekhar (1960) Chapter 1.

Bekefi (1966) Chapter 1

Ginzburg and Syrovatskii (1969)

Influence of Plasmas and Magnetoactive Effects

Ginzburg (1964)

Landau and Lifshitz (1960) Chapter 11

Gardner and Whiteoak (1966)

Problems

1.1. The redshift $z = \Delta\lambda/\lambda = 2$ for receding galaxies corresponds to what value of v/c?

1.2. Show that, for a source with redshift z, the observed flux density $F(\nu)$ is related to the emitted flux density $F'(\nu)$ by

$$F(\nu) = F'[(1+z)\nu]/(1+z)$$

and the total fluxes are related by

$$F = F'(1+z)^2$$

1.3. A quarter-wave plate and a polarization filter are placed along the path of a beam of monochromatic light. Before entering the quarter-wave plate, the light has right-handed elliptical polarization; the ratio of the major to the minor axes is 4:1. No light is transmitted through the polarization filter. Show in a diagram the orientation of the axes of the plate and of the transmission axis of the filter with respect to the axes of the ellipse. Compute the angle formed by the transmission axis of the filter with the y-axis.

1.4. Magnetic dipole radiation is described by the same formulas as electric dipole radiation, with the electric dipole moment replaced by the magnetic dipole moment, and the electric vector rotated by 90°. Compute the radiation from a rotating magnetic star in which the magnetic moment is perpendicular to the axis of rotation. In particular if the magnetic moment is M and the angular velocity of rotation is ω, show that:

(a) the angular distribution of the radiation averaged over the period of the rotation is

$$dP/d\Omega = M^2 \omega^4 (1+\cos^2 \theta)/8\pi c^3$$

where θ is the angle between the direction of observation and the axis of rotation;

(b) the total radiation is

$$P = 2M^2 \omega^4 /3c^3$$

(c) the radiation along the axis of rotation is circularly polarized.

1.5. In some pulsar models (see, e.g., P. Sturrock, Ap. J. 164, 529, (1971) electrons are accelerated to ultrarelativistic speeds in a narrow cone near the surface of a neutron star and move away from the star along magnetic field lines. Since the lines are curved, they will emit "curvature radiation" as discussed in Section 1.10. Assuming a dipole configuration and considering only small angles near the pole, find the total power emitted and the peak frequency for the radiation from an electron of energy γmc^2.

1.6. Show that, if the index of refraction of the interstellar medium can be described by equation (1-185), then an infinitely sharp pulse of radiation emitted by a pulsar at a distance R from the earth will be smeared out at the receiver over a time

$$\Delta t = (R\omega_p^2/c\omega^3)\Delta\omega \qquad \text{sec}$$

where $\Delta\omega$ is the bandwidth of the receiver. Assume that $\omega >> \omega_p$, and the density is constant between the source and the observer.

Chapter 2

BASIC FORMULAS FOR QUANTUM RADIATION PROCESSES

As discussed at the beginning of Chapter 1, the classical approximation applies only when the frequency ν of the emitted radiation is much less than W/h, where W is the energy of the radiating particle and h is Planck's constant. For most astrophysical applications the classical approximation can be used to treat such important processes as synchrotron radiation, electron scattering and low-frequency bremsstrahlung (see Chapters 3,4, and 5). However, there are many equally important processes such as line radiation, high frequency bremsstrahlung and the photoelectric effect, for which a quantum-mechanical description is needed. The basic concepts and formulas needed to calculate quantum radiation processes are summarized in this chapter.

2.1 Energy and Momentum of a Photon

The development of quantum mechanics was preceded by that of the quantum theory of radiation. For, although all problems relating to the propagation of light could be understood within the framework of the wave theory, a number of important phenomena relating to the emission and absorption of radiation remained unexplained. For instance, the energy spectrum of a black body derived on the basis of the wave theory was in contradiction with experiment.

The development of quantum theory started in 1901 with Planck's hypothesis that radiation is emitted and absorbed in finite amounts called quanta or photons. With this assumption, the black body spectrum was readily understood from thermodynamic arguments.

The energy of a photon is proportional to the frequency ν of the oscillations of the radiation field.

$$W = h\nu = \hbar\omega \qquad (2\text{-}1)$$

where $\hbar = h/2\pi$.

Later Einstein showed the necessity of assigning to the photon besides the energy, a momentum $p = W/c$ whose direction is given by the wave vector $\vec{k}$. Thus,

$$\vec{p} = \hbar\,\vec{k} \qquad (2\text{-}2)$$

Equations (2-1) and (2-2) are fundamental to the quantum theory of radiation and relate the energy W and momentum $\vec{p}$ of the photon to the frequency ν and wavelength λ of a monochromatic plane wave whose direction of propagation is given by the vector $\vec{k}$.

Planck's hypothesis that the exchange of energy and momentum between electrons, atoms, molecules, etc. and radiation occurs by the creation and annihilation of photons can now be expressed mathematically in terms of conservation laws.

Let W and $\vec{p}$ be the energy and momentum of the system before the collision with a photon of energy $h\nu$ and momentum $\hbar\vec{k}$, and W', P', $h\nu'$, $\hbar\vec{k}'$ the same quantities after the collision. The laws of conservation of energy and momentum become

$$h\nu + W = h\nu' + W' \qquad (2\text{-}3)$$

$$\hbar\vec{k} + \vec{p} = \hbar\vec{k}' + \vec{p}'$$

The case of $\nu' = 0$ refers to the absorption of a photon of energy $h\nu$; $\nu = 0$ to the emission of a photon of energy $h\nu'$. If ν and ν' are not zero, the equations define the scattering of radiation.

It is important to note that these laws are in conflict with both the wave and corpuscular concepts of radiation and cannot be interpreted within the framework of classical physics.

According to the wave theory the energy of a wave field is given by $(E^2 + B^2)/8\pi$, independent of the frequency ν. There is no general relation between the wave amplitude and the oscillation frequency which would allow the energy of a photon to be related to the wave amplitude. The assumption that a photon is a particle located somewhere in space is also invalid. A photon, by definition, is associated with a monochromatic plane wave. Such a wave is a purely periodic process, infinite in both space and time. The assumption that the photon is localized is in contradiction with the complete periodicity of the wave.

Thus if we accept Equation (2-1) through (2-4), radiation must have both wave and corpuscular properties.

These laws have been verified by a number of experiments. For example, in the photoelectric effect the velocity of the photo electrons depends solely on the frequency ν

of the light and not at all on the intensity of the incident light. These observations cannot be interpreted classically, but are easily understood in terms of the conservation laws (2-3) and (2-4).

Another example of the experimental confirmation of the conservation laws is the Compton effect. From Equations (2-3) and (2-4) it follows that the wavelength of the radiation scattered from an electron at rest is lengthened by an amount

$$\Delta\lambda = (2\ h/m_e c)\ \sin^2(\theta/2) \tag{2-5}$$

This result is in complete agreement with the observations.

The length $\lambda_c = h/m_e c = 2.4 \times 10^{-10}$ cm is called the Compton wavelength, and is one of the scales of the micro-universe.

According to the classical theory, which assumes continuity of the exchange of energy between the field and the microsystems, $h=0$ and no frequency shift should occur when light is scattered by a free electron (see 2-5). A direct calculation by the classical theory leads to the same result (see Chapter 4).

2.2 Elementary Quantum Theory of Radiation; Black Body Radiation

The elementary theory of radiation on the basis of quantum ideas is due to Einstein. It is to some extent phenomenological but his hypotheses are fully justified by modern quantum electrodynamics.

Consider two states of any system, denoting one by the letter m and the other by n. Let the energy of the first state be W_m and of the second be W_n. Assume $W_m > W_n$.

From experiment it is known that a system can spontaneously jump from a higher state to a lower one, emitting a photon with frequency

$$\nu = (W_m - W_n)/h \tag{2-6}$$

in the process. The photon has a definite polarization and a propagation vector $\vec{k}$ in a solid angle $d\Omega$. A polarization for a given direction of propagation of light can be represented as a superposition of two independent polarizations ℓ_1 and ℓ_2 in perpendicular directions. In the transition $\underline{m} \rightarrow \underline{n}$ a photon can be emitted with polarization either ℓ_1 or ℓ_2. Denote the polarization by the subscript α, and the probability per second of the transition $m \rightarrow n$ with emission of a photon of frequency $\nu = (W_m - W_n)/h$ into the solid angle $d\Omega$ by

$$dJ_s = A_{mn\alpha}\, d\Omega/4\pi \tag{2-7}$$

J_s is called the spontaneous transition probability.

Radiation incident on an atom can either be absorbed $(n \rightarrow m)$ with a probability dJ_a, or, if the atom is in an excited state, can induce a transition to a lower state $(m \rightarrow n)$ with a probability dJ_i. The probabilities for absorption and induced emission are proportional to the flux of incident photons. Both types of transition have analogues in the classical theory: an oscillator under the action of external radiation can either absorb or emit energy, depending on the relation between the phase of its oscillations and the phase of the light waves.

Denote the intensity of radiation propagated in directions within the solid angle $d\Omega$ with frequency between ν and $\nu + d\nu$ by $I(\nu,\Omega)d\nu d\Omega$. The Einstein differential coefficients are defined by the Equation (2-7) (we sum over polarization) and

$$dJ_a = B_{nm} \, I(\nu,\Omega)(d\Omega/4\pi)$$
$$dJ_i = B_{mn} \, I(\nu,\Omega)(d\Omega/4\pi) \qquad (2\text{-}8)$$

They depend only on the nature of the system and can be calculated by the methods of quantum mechanics (see Section 2.7). However, general relations between the coefficients can be derived from thermodynamic arguments.

Consider conditions in which there is equilibrium between emission and absorption. Let the number of atoms in the excited state m be N_m and the number in the lower state be N_n. In equilibrium the number of absorptions must be equal to the number of emissions:

$$N_m \, [A_{mn} + I(\nu,\Omega)B_{mn}] = N_n \, B_{nm} \, I(\nu,\Omega) \qquad (2\text{-}9)$$

The relative number of atoms in the states n and m is given by the Boltzmann formula:

$$N_n/N_m = (q_n/q_m) \, e^{(W_m - W_n)/KT} \qquad (2\text{-}10)$$

In equation (2-10), q_n and q_m are the <u>statistical weights</u> of the states n and m, T is the temperature and K is Boltzmann's constant. They give the relative number of atoms in the states n and m at infinite temperature. Substitution into (2-9) yields

$$B_{nm}\, I(\nu,\Omega) = (q_m/q_n)[A_{mn\alpha} + I(\nu,\Omega)B_{mn}] \cdot e^{(W_n-W_m)/KT} \qquad (2\text{-}11)$$

As the temperature $T \to \infty$, $I(\nu,\Omega) \to \infty$, so

$$B_{nm} = (q_m/q_n)B_{mn} \qquad (2\text{-}12)$$

Using this relation and (2-6), we find that the intensity in equilibrium is given by

$$I(\nu,\Omega,T) = (A_{mn}/B_{mn})/(e^{h\nu/KT} - 1) \qquad (2\text{-}13)$$

The relation between A_{mn} and B_{mn} can be determined by taking the classical limit. Consider the radiation in a cubical enclosure of side X. Assume that the fields vanish at the boundaries. The waves in the cavity will have the form (see equation (1-73))

$$f = f_o \exp\{\pm\, i(\vec{k}\cdot\vec{r} - \omega t)\} \qquad (2\text{-}14)$$

with

$$\vec{k} = (2\pi/X)(n_x\,\hat{i} + n_y\,\hat{j} + n_z\,\hat{k}) \qquad (2\text{-}15)$$

where n_x, n_y and n_z are positive integers, and

$$|\vec{k}| = (2\pi/X)(n_x^2 + n_y^2 + n_z^2)^{\frac{1}{2}} = 2\pi\nu/c \qquad (2\text{-}16)$$

According to the equipartition theorem of classical statistical mechanics, the average energy associated with each independent vibration of the field in equilibrium is KT. The energy density of the electromagnetic field under conditions of thermodynamic equilibrium is therefore determined by the number of independent vibrations having frequencies lying between ν and $\nu + d\nu$. This is equivalent to finding the number of vectors k whose lengths are between $2\pi\nu/c$ and $(2\pi/c)(\nu + d\nu)$, subject to the restriction given in Equation (2-16). The integers n_1, n_2 and n_3 can take on positive and negative values, so taking into account the two independent polarizations, there are 16 independent waves for each set of positive integral values of n_1, n_2, n_3. The number of combinations of integers to a frequency less than ν is just the number of points in the octant of a sphere of radius $n = \nu X/c$; namely $\pi(\nu X/c)^3/6$. The number of independent waves between frequencies ν and $\nu + d\nu$ is therefore $(8\pi X^3/c^3)\nu^2 \, d\nu$. Giving an energy KT (½ KT kinetic, ½ KT potential) to each one of these vibrations yields the classical Rayleigh-Jeans result for the spectral energy density.

$$U(\nu,T) = (8\pi/c^3)\,KT\nu^2 \qquad (2\text{-}17)$$

The energy density is related to the intensity by

$$U(\nu) = (1/c) \int I \, d\Omega \qquad (2\text{-}18)$$

So for the isotropic equilibrium radiation field

$$I = 2\,KT\nu^2/c^2 \qquad (2\text{-}19)$$

According to the correspondence principle the classical and quantum theories should yield the same result for large quantum numbers. In this case the quantum number $n = KT/h\nu$, so the two theories should agree when n is much greater than unity, or what is the same thing, when $h\nu \ll KT$. In this limit (2-13) becomes

$$I = (A_{mn}/B_{mn})(KT/h\nu) \qquad (2\text{-}13')$$

Therefore

$$A_{mn}/B_{mn} = 2\, h\nu^3/c^2 \qquad (2\text{-}20)$$

and the intensity in equilibrium is

$$I_{eq}(\nu) \equiv B(\nu) = (2\, h\nu^3/c^2)/(e^{h\nu/KT} - 1) \qquad (2\text{-}21)$$

This is Planck's Law for the intensity of a radiation field in thermodynamic equilibrium with its surroundings; such a radiation field is called black-body radiation. The Planck function (2-21) reaches a maximum at the frequency defined by

$$h\nu_m = 2.82\ KT \qquad (2\text{-}22)$$

or

$$\nu_m = 5.88\times10^{10}\ T \quad hz \qquad (2\text{-}22')$$

The mean frequency of the radiation is given by

$$h\bar{\nu} = 3.83 \ KT \qquad (2\text{-}23)$$

or

$$\bar{\nu} = 7.97\times 10^{10} \ T \quad hz \qquad (2\text{-}23')$$

In terms of the wavelength $\lambda = c/\nu$, the Planck function takes the form

$$B(\lambda)d\lambda = (2hc/\lambda^5)d\lambda/(e^{hc/KT\lambda} - 1) \qquad (2\text{-}24)$$

The distribution of the energy density with respect to wavelength has a maximum at a wavelength given by

$$hc/\lambda_m = 4.97 \ KT \qquad (2\text{-}25)$$

or

$$\lambda_m = 2.90\times 10^7/T \quad \text{Å} \qquad (2\text{-}25')$$

The intensity of black-body radiation integrated over all frequencies is

$$I_{eq} = \int_0^\infty B(\nu)d\nu = (2\ K^4 T^4/h^3 c^2)\int_0^\infty x^3(e^x-1)^{-1}\ dx \qquad (2\text{-}26)$$

The integral is equal to $\pi^4/15$, so

$$I_{eq} = (\sigma/\pi)T^4 \qquad (2\text{-}27)$$

where the <u>Stefan-Boltzmann constant</u> σ is given by

$$\sigma = 2\pi^5 K^4/15\ h^3 c^2 = 5.67\times 10^{-5} \ gm/sec\ deg^4 \qquad (2\text{-}28)$$

The energy density of the radiation is

$$U = (4\pi I/c) = (4\sigma/c)T^4 \qquad (2\text{-}29)$$

The density of flow of energy coming from a direction $\hat{n}$ lying in the element of solid angle $d\Omega$ is $cI d\Omega$. The flux of radiation F across a unit area of a surface at an angle θ to its normal is

$$dF/d\Omega = I \cos\theta \qquad (2\text{-}30)$$

The total flux from a black body is obtained by integrating (2-30) over all solid angles in a hemisphere:

$$F = \int_0^{\pi/2} I \cos\theta \; 2\pi \sin\theta \; d\theta = cU/4 = \sigma T^4 \qquad (2\text{-}31)$$

The spectral distribution of the intensity and/of the flux are obtained simply by replacing U by $U(\nu)$.

For radiation not in thermal equilibrium, the spectral distribution will not be Planckian and the intensity may not be isotropic. The "brightness temperature" T_B of a nonthermal radiation field may be defined as the temperature for which the intensity $I(\nu,\hat{n})$ in the direction $\hat{n}$ is equal to the value given by the Planck formula.

In the classical limit ($h\nu \ll KT$) the brightness temperature is given by

$$KT_B(\hat{n}) = c^2 \; I(\nu,\hat{n})/2\nu^2 \qquad (2\text{-}32)$$

The brightness temperature is sometimes, though not always, equal to the <u>antenna tem-</u>

perature, which is defined in terms of the power absorbed by the antenna. If $A(\hat{n})$ is the effective area of the antenna for absorbing radiation from a direction $\hat{n}$, then the antenna temperature T_a is given by (Bekefi, 1966)

$$T_a = (1/2K)\int I(\nu,\hat{n})A(\hat{n})d\Omega$$

$$T_a = (1/\lambda^2)\int T_B(\hat{n})A(\hat{n})d\Omega \;/\; \int A(\hat{n})d\Omega \qquad (2\text{-}33)$$

Equation (2-33) shows that for an extended source whose brightness temperature does not vary over the main antenna lobe, $T_a = T_b$.

2.3 The Schrodinger Equation

In the classical limit (see Introduction to Chapter 1) the position and momentum of an electron can be specified without any uncertainty at each point of space and at every moment. Once the external fields are specified all the radiation processes can in principle be computed by using Newton's second law and Maxwell's equations. In the quantum limit we must modify Newton's second law and quantize the radiation field to take into account the dual wave-particle nature of electrons and photons. The first part of this task is fairly easily accomplished by the introduction of the wave function Ψ which can be used to find the probability that a particle is at a given place with a given momentum at a given time, and by replacing the equation of motion by a differential equation for the wave function.

The quantization of the radiation field is a task which is beyond the scope of this book. Fortunately most of the results quoted in this book can be derived by a semi-classical treatment in which the motion of the particle is treated quantum mechanically, but the radiation field is treated classically, except that the wave-particle nature of photons is taken into account by means of the simple theory discussed in Sections 2.1 and 2.2.

The equation for the wave function (wave equation) is related to the classical equation of motion in the following way. The Lorentz equation can be written in terms of Hamilton's equations (Kramers, 1958, or almost any other text on quantum mechanics):

$$dP_k/dt = -\,\partial H/\partial q_k \quad ; \quad dq_k/dt = \partial H/\partial P_k \qquad (2\text{-}33)$$

where P_k and q_k are the canonically conjugate momenta and coördinates, and H is the Hamiltonian function. For an electron in an electromagnetic field (Landau and Lifshitz, 1962)

$$P_1 = p_x - (e/c)A_x \quad ; \quad q_1 = x \; , \; q_2 = y \; , \text{ etc.} \qquad (2\text{-}34)$$

where p_x is the component of the momentum of the particle in the x-direction, A_x is the x-component of the vector potential, etc. The Hamiltonian function H is equal to the total energy of the electron including the rest mass energy:

$$H = -\,e\varphi + [m_e^2c^4 + c^2(\vec{P} + (e/c)\vec{A})^2]^{1/2} = W \qquad (2\text{-}35)$$

In the non-relativistic limit, the Hamiltonian is

$$H = m_e c^2 - e\varphi + P^2/2m_e - (e/2m_e c)(\vec{P}\circ\vec{A} + \vec{A}\cdot\vec{P})$$
$$+ (e^2 A^2/2m_e c^2) \qquad (2\text{-}36)$$

The non-relativistic wave equation is obtained by setting $(H - m_e c^2)\Psi = W\Psi$ and replacing the conjugate momentum and the energy by the following differential operators:

$$\vec{P} \rightarrow -i\hbar\vec{\nabla} \quad ; \quad W \rightarrow i\hbar\frac{\partial}{\partial t} \qquad (2\text{-}37)$$

where $\hbar = h/2\pi$.

The result is

$$-(\hbar/2m_e)\nabla^2\Psi - (e\hbar/2i\,m_e c)(\vec{A}\cdot\vec{\nabla} + \vec{\nabla}\cdot\vec{A})\Psi \qquad (2\text{-}38)$$
$$+ (e^2/2m_e c^2)A^2\Psi - e\varphi\Psi = i\hbar\,\partial\Psi/\partial t$$

The wave equation in this form is called the Schrödinger equation for an electron or charge -e in an electromagnetic field described by a vector potential A and a scalar potential φ. It is valid for a single non-relativistic electron with no spin. The effects of spin are discussed in Section 2.5

2.4. Characteristic States, or Eigenstates

The wave function represents the state of the system in the sense that the probability that a measurement of the various coordinates x, y, z will give results lying in the range x to (x + dx), etc. at the time t is given by

$$dw(x,y,z,t) = |\Psi(x,y,z,t)|^2\,dxdydz \qquad (2\text{-}39)$$

If w is to define the probability in the

usual sense of the word, then the wave function must satisfy the normalization requirement

$$\int |\Psi(x,y,z,t)|^2 \, dxdydz = 1 \qquad (2\text{-}40)$$

where the integral is over all values of the coordinates.

The average value at time t of any dynamical quantity Q is given by

$$Q = \int \Psi^* \, \tilde{Q} \, \Psi \, dxdyz \qquad (2\text{-}41)$$

where Ψ^* is the complex conjugate of Ψ and $\tilde{Q}$ is the operator that represents Q. The operator that represents the coordinate is "multiplication by x" so

$$\overline{x} = \int \Psi^* \, x \, \Psi \, dxdydz \qquad (2\text{-}42)$$

The operator that represents the momentum is $- i\hbar\vec{\nabla}$ so

$$\overline{\vec{p}} = -i\hbar \int \Psi^* \, \vec{\nabla} \, \Psi \, dxdydz \qquad (2\text{-}43)$$

The operator for any function of the coordinates and the momentum is formed by replacing the conjugate momentum $\vec{P}$ by $- i\hbar\vec{\nabla}$ while leaving the coordinates unchanged.

When a system is in a state such that some dynamical quantity has a precisely defined value, the state is said to be characteristic of that dynamical quantity. The function which describes the state is called a characteristic function or eigenfunction of the operator corresponding to the dynamical quantity. The precisely defined value is called a characteristic value or eigenvalue.

The general equation for a characteristic function is

$$Q\chi = Q_o\chi \qquad (2\text{-}44)$$

where Q is the operator corresponding to the dynamical quantity of which χ is the eigenfunction. The eigenvalue of the operator Q is Q_o when the system is in the state defined by eigenstates which have the same eigenvalues are said to be degenerate.

In general there are only certain values of Q_o for which equation (2-44) is satisfied. This means that there are only certain values of the dynamical quantity Q which can be determined experimentally. For example, the quantum theory of radiation is based on the assumption that the normal vibrations of the electromagnetic field can have only discrete values of the energy: $W = nh\nu$, where n is an integer.

For a one-dimensional system the function $\delta(x - x_o)$ is characteristic of the quantity x and represents a state in which x has the value x_o. Similarly, the function $e^{(i/h)p_o x}$ represents a state characteristic of the momentum in which the momentum has the value p_o.

States in which an isolated physical system has a definite energy are particularly important. Actual atomic systems and other systems are not quite isolated because they are coupled to the electromagnetic field. For many problems this coupling is weak and the description of the system in terms of energy states is useful as a close approximation.

The operator representing the energy is the Hamiltonian H, so the functions representing the energy states satisfy the equation

$$i \hbar \, \partial\Psi/\partial t = H\Psi = W\Psi \tag{2-45}$$

The equality of the first and last terms shows that

$$\Psi(x,y,z,t) = u(x,y,z)e^{-(i/\hbar)W_n t} \tag{2-46}$$

where u(x,y,z) is a function of the coordinates but not the time. The function u(x,y,z) also satisfies the second equality in (2-47). Hence for a single electron moving in an electric field, which is constant in time and defined by the potential φ, the eigenstates are obtained from the solutions to the time-independent Schrödinger equation:

$$Hu = -(\hbar^2/2m)\nabla^2 u - e\varphi u = Wu \tag{2-48}$$

Energy eigenstates are often called <u>stationary states</u> because the probability distribution $\Psi^*\Psi$ is constant in time, and the probability distributions of all quantities are constant.

A general solution of the Schrödinger equation (2-48) can be written in terms of energy eigenstates. The complete set of eigenfunctions $u_k(x,y,z)$ associated with the energies W_k form a complete orthogonal set, and as such satisfy the condition

$$\int u_k^* \, u_m \, dxdydz = \delta_{km} \tag{2-49}$$

This property enables us to expand the wave function in terms of the u_k:

$$\Psi\ (x,y,z,t) = \sum_k a_k u_k (x,y,z) e^{-(i/\hbar) W_k t} \qquad (2\text{-}50)$$

where the coefficients a_k are determined by the conditions of the problem at hand.

In only a few cases can the Schrödinger equation be solved exactly. One such case, the motion of a particle in a Coulomb field, is discussed in Chapter 6. For other problems approximation methods must be used. For radiation problems the Hamiltonian depends on the time through the radiation field, so we must use time-dependent perturbation theory.

2.5 Electron Spin

By 1925 it was clear that the description of an electron in terms of wave functions which depend only on the coordinates of position and time was inadequate. The Stern-Gerlach experiment, the Zeeman effect and the multiple structure of atomic spectra all presented difficulties which could be reconciled only if it was assumed that the electron had an intrinsic magnetic moment. In particular, Uhlenbeck and Goudsmit showed that the observations would be explained by assuming that the electron behaves as a charged particle which is rotating around an axis in such a way that it possesses an angular momentum of absolute magnitude $\frac{1}{2}\hbar$ around that axis and that it possesses a magnetic moment $e\hbar/2mc$ oppositely directed to the angular momentum. The existence of only two characteristic values ($\pm \frac{1}{2} \hbar$) for the spin angular momentum distinguishes it from ordinary angular momentum. No measurement will ever show a component less in magnitude than $\hbar/2$.

The spin operator cannot be constructed, as in the case of other operators, by analogy with classical mechanics, because it is essentially a quantum-mechanical quantity that does not depend on the coordinates and the momenta and it vanishes in the classical limit $\hbar \to 0$. However, the Uhlenbeck-Goudsmit model suggests an analogy with the classical angular momentum which can be used to construct the spin operators. First, we review certain properties of the orbital angular momentum operator.

Classically the orbital angular momentum is

$$L = \vec{r} \times \vec{p} \tag{2-51}$$

Following the prescription given in Section 2.4, angular momentum operator is

$$\vec{L} = - ir \times \hbar\vec{\nabla} = - i\hbar \begin{vmatrix} i & j & k \\ x & y & z \\ \frac{\partial}{\partial x} & \frac{\partial}{\partial y} & \frac{\partial}{\partial z} \end{vmatrix} \tag{2-52}$$

In particular the z-component is

$$L_z = - i\hbar \left(x \frac{\partial}{\partial y} - y \frac{\partial}{\partial x}\right) = \frac{\partial}{\partial \varphi} \tag{2-53}$$

In spherical polar coordinates

$$L_z = (\hbar/i)(\partial/\partial\varphi) \tag{2-54}$$

and functions characteristic of L_z are

$$v = f(r,\theta)e^{im\varphi} \tag{2-55}$$

The commutation relations for $\vec{L}$ are

$$(\vec{L} \times \vec{L})\psi = - i \hbar \vec{L}\psi \tag{2-56}$$

The fact that the operators for the different components of the angular momentum do not commute means that the precise specification of one component precludes the precise specification or the others. The operator for each component does commute with the operator for L^2.

$$L^2 \vec{L} \psi = \vec{L} L^2 \psi \tag{2-57}$$

In spherical polar coordinates L^2 is given by

$$L^2 = - \hbar^2 \left\{\frac{1}{\sin\theta} \frac{\partial}{\partial\theta}(\sin\theta \frac{\partial}{\partial\theta}) + \frac{1}{\sin^2\theta} \frac{\partial^2}{\partial\varphi^2}\right\} \tag{2-58}$$

The solution to the characteristic equation for L^2

$$L^2 Y = L_o^2 Y \tag{2-59}$$

is

$$Y = P_\ell^m(\cos\theta)\ e^{im\varphi} \tag{2-60}$$

where $P_{\ell}^{m}(\cos\theta)$ is an associated Legendre polynomial. In order for it to be single valued, definite and quadratically integral in the whole range of the variable θ, m and ℓ must satisfy the restrictions.

$$\begin{aligned} &m, \ell \text{ integers} \\ &\ell \geq m \\ &\ell(\ell+1) = L_o^2/\hbar^2 \end{aligned} \tag{2-61}$$

Returning now to the consideration of the electron spin operator, we require that the operators for the 3 components of the vector spin obey the same commutation rules as those for the angular momentum. Thus, if s_x, s_y, s_z are the operators, then

$$\begin{aligned} s_x s_y - s_y s_x &= i\, s_z \\ s_z s_x - s_x s_z &= i\, s_y \\ s_y s_z - s_z s_y &= i\, s_x \end{aligned} \tag{2-62}$$

The characteristic values of s_x, s_y, s_z are $\pm \frac{1}{2}$ (in units of $\hbar$) so

$$s_x^2 = s_y^2 = s_z^2 = (\tfrac{1}{4})I \tag{2-63}$$

where I is the identity operator ($I\psi = \psi$). The commutation rules show that it is not possible to exactly specify all three components of $\vec{s}$ at the same time. However, any one component, say the z-component, can be fixed. The two possible values of the spin are $\pm \frac{1}{2}$. They are the values which the argument of the spin eigenfunction S(s) is normalized to unity then

$$\sum_{\alpha}^{1} |S(s\alpha)|^2 = |S(+\tfrac{1}{2})|^2 + |S(-\tfrac{1}{2})|^2 = 1 \quad (2\text{-}64)$$

If two different spin functions S_1 and S_2 are orthogonal to each other, then

$$S_1^*(\tfrac{1}{2})S_2(\tfrac{1}{2}) + S_1^*(-\tfrac{1}{2})S_2(-\tfrac{1}{2}) = 0 \quad (2\text{-}65)$$

If $|S(+\frac{1}{2})|^2 = 1$ and $S(-\frac{1}{2}) = 0$, then the function S represents a state in which the spin in the z-direction is fixed at $+\frac{1}{2}$. In general $|S(+\frac{1}{2})|^2$ is the probability that the spin has the value $+\frac{1}{2}$, and $|S(-\frac{1}{2})|^2$ is the probability that the spin is $-\frac{1}{2}$. Thus the spin function is to be interpreted in the same way as a wave function having the coordinates as variables, with the difference that the argument of the spin function can take only two values, so that integration is replaced by summation.

The wave function for an electron must now be written as a function of the four coordinates x,y,z,s, and the time.

$$\psi = \psi(x,y,z,s,t) = \begin{matrix} \psi(x,y,z + \tfrac{1}{2},t) \\ \psi(x,y,z - \tfrac{1}{2},t) \end{matrix} \quad (2\text{-}66)$$

$|\psi(x,y,z,\frac{1}{2},t)|^2$ is the probability that the electron has the coordinates x,y,z with the spin of $\frac{1}{2}$ in the z-direction at the time t, etc. If the space motion and the spin are strongly coupled, then the functions $\psi(x,y,z,\frac{1}{2},t)$ and $\psi(x,y,z,-\frac{1}{2},t)$ will be quite different. If the coupling can be ignored, then

$$\psi(x,y,z,s,t) = \psi(x,y,z,t)\ S(s) \tag{2-67}$$

The operator s_z represents an argument of the wave function, so it must be analogous to the coordinate operator. It must multiply $\psi(x,y,z,\frac{1}{2},t)$ by $\frac{1}{2}$ and $\psi(x,y,z,-\frac{1}{2},t)$ by $-\frac{1}{2}$. If we consider the pair of wave functions $\psi(x,y,z,+\frac{1}{2},t)$ and $\psi(x,y,z,-\frac{1}{2},t)$ as components of a vector, then the spin operators can be represented by matrices. The matrix for s_z is obviously

$$s_z = \tfrac{1}{2}\begin{pmatrix} 1 & 0 \\ 0 & -1 \end{pmatrix} \tag{2-68}$$

With this choice for s_z, the commutation rules fix s_x and s_y:

$$s_x = \tfrac{1}{2}\begin{pmatrix} 0 & 1 \\ 1 & 0 \end{pmatrix} \qquad s_y = \tfrac{1}{2}\begin{pmatrix} 0 & -i \\ i & 0 \end{pmatrix} \tag{2-69}$$

Writing $\vec{s} = \frac{1}{2}\vec{\sigma}$ defines the <u>Pauli spin matrices</u>:

$$\sigma_x = \begin{pmatrix} 0 & 1 \\ 1 & 0 \end{pmatrix} \qquad \sigma_y = \begin{pmatrix} 0 & -i \\ i & 0 \end{pmatrix} \qquad \sigma_z = \begin{pmatrix} 1 & 0 \\ 0 & -1 \end{pmatrix} \tag{2-70}$$

The eigenvalue equation for $S(s)$ is

$$s_z\ S(s) = m_s\ S(s) \tag{2-71}$$

where $m_s = \pm \frac{1}{2}$. The characteristic functions are therefore

$$S(+\tfrac{1}{2}) \equiv \alpha = \begin{pmatrix} 1 \\ 0 \end{pmatrix} \qquad S(-\tfrac{1}{2}) \equiv \beta = \begin{pmatrix} 0 \\ 1 \end{pmatrix} \tag{2-72}$$

For a system consisting of two electrons, the spin operator is $S = s_1 + s_2$, where s_1 operates only on the first electron and s_2 only on the second. The eigenvalues of the spin in the z-direction are now 1, 0 and -1. The eigenvectors for S_z and S^2 are

$$\alpha_1 \alpha_2 \quad \text{corresponding to } S=1, M_s=+1$$

$$1/\sqrt{2}\,(\alpha_1 \beta_2 + \beta_1 \alpha_2) \quad \text{corresponding to } S=1, M_s=0$$

$$\beta_1 \beta_2 \quad \text{corresponding to } S=1, M_s=-1$$

$$1/\sqrt{2}\,(\alpha_1 \beta_2 - \beta_1 \alpha_2) \quad \text{corresponding to } S=0, M_s=0$$

(2-73)

The first three states, correspondent to S=1 are described by symmetric spin functions. They give rise to triplet structure in atomic spectra. The state with S=0 is anti-symmetric; it gives rise to singlet lines.

Hamiltonian Term For The Spin in a Magnetic Field. Because of the magnetic moment of the electron there is an interaction energy between the spin and any magnetic field present. The part of the Hamiltonian which represents this interaction can be written as

$$H_s = -(e\hbar/m_e c)\,\vec{S} \cdot \vec{B} \qquad (2\text{-}74)$$

Spin-Orbit Interaction. In addition to any external magnetic fields which may be present, there will also be a magnetic field generated by the motion of the electron. The interaction of the magnetic moment of the electron with this field is called the spin-orbit interaction. In the non-relativistic approximation the term in the Hamiltonian corresponding to the spin-orbit interaction is (see Bethe and Jackiw 1968)

$$H_{so} = -(e\hbar/2m_e c^3)(r^{-1}\, d\Phi/dr)\,\vec{S}\cdot\vec{L} \qquad (2\text{-}75)$$

2.6 Time-Dependent Perturbation Theory; Transition Probabilities

Consider a system with an energy operator H which can be split into an unperturbed part and a small perturbing part:

$$H = H^{o} + H' \qquad (2\text{-}76)$$

In the case of radiation problems where equation (2-38) can be used, the unperturbed part is

$$H^{o} = -\ (\hbar^2/2m_e)\nabla^2 - e\varphi \qquad (2\text{-}77)$$

where φ is the potential for some constant electric field. The perturbing part is composed of those terms containing the first order terms in the vector potential A. The inclusion the effects of the electron spin will not change the results derived in this section, since the spin terms are almost always treated as second order terms.

We assume that H^{o} does not contain the time explicitly and that we know the eigenvalues W_k and the orthonormal set of eigenfunctions u_k. Degeneracy is not introduced explicitly so that several W_k may be the same. The procedure is to transform the

Schrodinger equation in such a way that the physical situation is described in terms of the probability amplitude $a_k(t)$ with respect to the eigenfunctions of the operator H^o.

First, we expand the wave function in terms of the u_k:

$$\Psi(t) = \sum_k a_k u_k e^{-(i/\hbar)W_k t} \qquad (2\text{-}78)$$

We then multiply equation (2-78) by u_k^* and integrate. Because of the orthonormal property (2-49) of the u_k, only one term in the summation survives, and we obtain:

$$\mathring{a}_k = e^{(i/\hbar)W_k t} \int u_k^* \Psi \, dxdydz \qquad (2\text{-}79)$$

Likewise, if we multiply (2-78) by $\Psi^* H^o$ and integrate, we find

$$\int \Psi^* H^o \Psi \, dxdydz = \sum_k |a_k(t)|^2 \, W_k \qquad (2\text{-}80)$$

The left hand side of (2-80) is the average value of H^o, whereas the right hand side is the sum of terms, each of which is a possible value of H^o multiplied by a term which must be the probability of that given value of H^o. Thus $|a_k(t)|^2$ is the probability that the system is in the state k at time t. If the system was in a state described by u_i initially, then the transition probability from state i to state k at time t is

$$G_{ki} = |a_k(t)|^2 \qquad (2\text{-}81)$$

Thus the problem is to determine $a_k(t)$ from the wave equation with the initial condition

$$a_k(0) = \delta_{ki} \tag{2-82}$$

Using equations (2-76) and (2-78) we can write the Schrodinger equation in the form

$$\sum_k a_k(W_k + H')u_k\, e^{-(i/\hbar)W_k t}$$

$$= \sum_k (i\hbar\, \mathring{a}_k + a_k W_k)\, e^{-(i/\hbar)W_k t} \tag{2-83}$$

We then multiply both sides of equation (2-83) by $u_f^* \, e^{(i/\hbar)W_k t}$ and integrate over the volume to obtain

$$i\hbar\, \dot{a}_f = \sum_k a_k (H')_{fk} \exp(i\, \omega_{fk} t) \tag{2-84}$$

where

$$\omega_{fk} = (W_f - W_k)/\hbar \tag{2-85}$$

and

$$(H')_{fk} = \int u_f^* \, H' \, u_k \, dxdydz \tag{2-86}$$

and it is assumed that H' commutes with $\exp(-\, iW_k t/\hbar)$.

To solve equation (2-84) we assume that $a_k(t)$ can be written as the sum of two parts, a zero-order term, and a first order term which is related to the perturbing part of the Hamiltonian:

$$a_k(t) = a_k^o(t) + a_k'(t) \qquad (2\text{-}87)$$

Equation (2-84) then becomes

$$i\hbar \dot{a}_f^o + i\hbar \dot{a}_f' = \sum_k (a_k^o + a_k')(H')_{fk} \exp(i\omega_{fk} t) \qquad (2\text{-}88)$$

All the terms on the right hand side are first order, so the zero-order equation is

$$\dot{a}_f^o = 0 \qquad (2\text{-}89)$$

With the initial condition (2-82), the solution for a_k^o is

$$a_k^o(t) = \delta_{ki} \qquad (2\text{-}90)$$

Substituting this solution into equation (2-88), the equation for the first order terms becomes

$$i\hbar \dot{a}_f' = (H')_{fi} \exp(i\omega_{fi} t) \qquad (2\text{-}91)$$

The solution for a_f' is therefore

$$a_f' = -(i/\hbar) \int_0^\tau e^{i\omega_{fi} t'} (H'(t'))_{fi}\, dt' \qquad (2\text{-}92)$$

Assume that the perturbation is zero for $t < 0$ and $t > \tau$. Then

$$a'_f = -(i/\hbar)\int_{-\infty}^{\infty} (H'(t'))_{fi}\, e^{i\omega_{fi}t'}\, dt' \quad (2\text{-}93)$$

For $t > \tau$, a'_f is independent of time since the energy is an integral of the motion. The Fourier components of $(H'(t'))_{fi}$ are given by

$$(H'(\omega))_{fi} = (1/2\pi)\int_{\infty}^{\infty} (H'(t'))_{fi}\, e^{i\omega t'}\, dt' \quad (2\text{-}94)$$

Hence

$$a'_f = (2\pi/i\hbar)(H'(\omega_{fi}))_{fi} \quad (2\text{-}95)$$

and the transition probability from the state i to the state f is (see equation (2-81))

$$G_{fi} = (4\pi^2/\hbar^2)\,|(H'(\omega_{fi}))_{fi}|^2 \quad (2\text{-}96)$$

The transition probability exhibits a resonance behavior, i.e., it is non-zero only for perturbations which contain frequencies equal to the eigenfrequencies of the system. For example, in the case of absorption of radiation by an atom, only those photons having an energy equal to $W_f - W_i$ will be absorbed.

For many problems H' can be assumed to be constant during the time of interaction. That is, the characteristic time τ for the interaction is much greater than the characteristic time scale of the system which is $\sim \omega_{fi}^{-1}$. Then equation (2-92) can be integrated to yield

$$a'_f = -(1/W_f - W_i)\,(\exp(i\omega_{fi}\tau) - 1))(H')_{fi} \tag{2-92'}$$

$$|a'_f|^2 = 4(H')^2_{fi}\,\sin^2(\omega_{fi}\tau/2)/(W_f - W_i)^2 \tag{2-93'}$$

(Remember $W_f - W_i = \hbar\,\omega_{fi}$.)

The transition probability per unit time is

$$w_{fi} = |a'_f|^2/\tau = (2\pi/\hbar)(H')^2_{fi}\;\delta(W_f - W_i) \tag{2-96'}$$

since $\sin^2 at/\pi a t \to \delta(a)$ as $t \to \infty$.

We can make an order of magnitude estimate of the conditions for the validity of the first order perturbation theory for radiation problems. The ratio of the two terms in (2-36) or (2-38) which contain A is

$$(epA/m_e c)/(e^2A^2/2m_e c^2) = (2pc/eA) \tag{2-97}$$

Most radiation problems involve the interaction of a charge with the Coulomb field of another charge, so $e^2/b \sim W$, the kinetic energy of the particle (b is the distance of closest approach). In radiation problems where the quantum theory is needed, most of the kinetic energy of the charge is radiated away in a photon of energy $W = h\nu$. Thus $e^2/b \sim h\nu$, and $e \sim (bh\nu)^{\frac{1}{2}}$. From the uncertainty principle, the momentum is related to b by $p \sim h/b$. The vector potential of the radiation field is related to the electric vector E by $A \sim (c/\nu)E$, where ν is the frequency of the radiation. The energy density of the field is $\sim E^2/4\pi$. This is related to the average number density of photons n_{ph} by $E^2/4\pi \sim n_{ph}h\nu$, so $E \sim (4\pi n_{ph} h\nu)^{\frac{1}{2}}$, and $A \sim (c/\nu)(4\pi n_{ph}h\nu)^{\frac{1}{2}}$. Putting all these estimates together, the ratio of the terms containing A is

$$2pc/eA \sim 1/(n_{ph}b^3)^{\frac{1}{2}} \qquad (2\text{-}97')$$

For first order perturbation theory to apply, we must have $n_{ph}b^3 \ll 1$. That is, the average number of photons in the volume where the interaction occurs must be much less than one. If the reverse were true, there would be many photons present in the emitting region. Thus we say that keeping only terms linear in the vector potential corresponds to considering only cases where a single photon is emitted or absorbed by the system. For the emission or absorption of two or more photons the quadratic term in A must be used.

On the other hand, we are sure that the classical description of the electromagnetic field is valid if there are many photons in a sphere whose radius is equal to the wavelength of the radiation; i.e., if

$$n_{ph}\ \lambda^3 \gg 1 \qquad (2\text{-}98)$$

Using the relations $\lambda = c/\nu$, $W = h\nu$, $W \cong pv$, and $p \sim h/b$, the condition (2-98) can be written as

$$n_{ph}\ b^3(c/v)^3 \gg 1 \qquad (2\text{-}98')$$

Equations (2-97′) and (2-98′) define the conditions for the applicability of the semi-classical first-order radiation theory. Actually due to the linear dependence of the perturbation on A, the results derived for radiation fields sufficiently intense to satisfy (2-98′) must also be valid for weaker radiation fields, so the range of validity of the semi-classical is not in practice restricted by equation (2-98′).

These considerations do not hold for spontaneous emission. Experiment shows that an atom will spontaneously emit radiation regardless of whether an external field is present. A similar situation is encountered in classical mechanics. According to classical mechanics, a charged particle moving in a static magnetic field will not lose energy. If however, we take into account the fact that the moving charge produces an electromagnetic field which acts on it, we find that the particle will in fact lose energy and emit light. The problem of emission with allowance for reaction is not linear in the external field and the correspondence principle can not be extrapolated in a simple way to the spontaneous emission of one quantum. For a completely satisfactory theory, the electromagnetic field must be quantized. We will avoid the difficulty by postulating, in accordance with Einstein's theory, that spontaneous emission exists and is related to absorption in the manner specified by that theory. The transition probabilities obtained in this way are in complete agreement with the results of the quantum theory of radiation.

2.7. Absorption and Emission Probabilities

Absorption. To treat the problem of absorption of the external radiation or the induced emission of radiation in the presence of an external field we use the perturbation (see equation (2-38)).

$$H' = (i\, e\, \hbar/2m_e c)\, (\vec{A}\circ\vec{\nabla} + \vec{\nabla}\cdot\vec{A}) \tag{2-99}$$

The factor $\vec{\nabla}\circ\vec{A}\,\psi = \vec{A}\cdot\vec{\nabla}\psi + \psi(\vec{\nabla}\cdot\vec{A})$. If we use the Contomb gauge to describe the radiation field, then $\vec{\nabla}\cdot\vec{A} = 0$, and $\vec{\nabla}\cdot\vec{A}\psi = \vec{A}\cdot\vec{\nabla}\psi$. The perturbing Hamiltonian H' becomes

$$H' = -\, (i\, e\, \hbar/m_e c)(\vec{A}\cdot\vec{\nabla}) \tag{2-99'}$$

Assume that A is zero for $t < 0$ and for $t > \tau$ where τ is much longer than the period of oscillation of the waves in the radiation field. Then (2-96) applies and the problem becomes one of calculating $(H'(\omega_{fi}))_{fi}$:

$$(H'(\omega_{fi}))_{fi} = \frac{ie\hbar}{2\pi m_e c} \int_{-\infty}^{\infty} \vec{A}\cdot(e^{i\vec{k}\cdot\vec{r}}\vec{\nabla})_{fi}\, e^{i\omega_{fi}t}\, dt \tag{2-100}$$

where I have set $\vec{A} = \vec{A}(t)\exp(i\vec{k}\cdot\vec{R})$, and

$$(e^{i\vec{k}\cdot\vec{r}}\,\vec{\nabla})_{fi} = \int u_f^*\, e^{i\vec{k}\circ\vec{r}}\, \vec{\nabla}\, u_i \; dxdydz \tag{2-101}$$

which is independent of time. Since (Equation (2-43)) $\vec{\nabla} = i\vec{p}/\hbar$, we have

$$(e^{i\vec{k}\cdot\vec{r}}\,\vec{\nabla})_{fi} = (i/\hbar)(e^{i\vec{k}\cdot\vec{r}}\,\vec{p})_{fi} \tag{2-101'}$$

Therefore

$$(H'(\omega_{fi}))_{fi} = \vec{A}(\omega_{fi})(e/m_e c)(e^{i\vec{k}\cdot\vec{r}}\,\vec{p})_{fi} \quad (2\text{-}102)$$

and

$$G_{fi} = (2\pi/\hbar)^2 (e/m_e c)^2 \,|(e^{i\vec{k}\cdot\vec{r}}\hat{\ell}\cdot\vec{p})_{fi}|^2\, |A(\omega_{fi})|^2 \quad (2\text{-}103)$$

where $\hat{\ell}$ is a unit vector specifying the polarization of the wave. It is convenient to define the quantity

$$(\tilde{p})_{fi} \equiv (e^{\pm i\vec{k}\cdot\vec{r}}\,\hat{\ell}\cdot\vec{p})_{fi} \quad (2\text{-}104)$$

where the + sign in the exponential will be used when referring to absorption and the - sign when referring to emission. Now expand the vector potential in plane waves:

$$\vec{A} = \tfrac{1}{2}\int\{\vec{A}_o(\omega)e^{ik\cdot r-i\omega t} + \vec{A}_o^*(\omega)e^{-i\vec{k}\cdot\vec{r}+i\omega t}\}\,d\omega \quad (2\text{-}105)$$

where we have written $\vec{A}$ such that it is manifestly real. The Fourier transform of $\vec{A}$ is

$$\vec{A}(\omega_{fi}) = \vec{A}_o(\omega_{fi})\, e^{i\vec{k}_{fi}\cdot\vec{r}}/2 \quad (2\text{-}106)$$

where we have assumed that $\omega_{fi} > 0$, corresponding to absorption. Substitution into (2-103) yields the transition probability:

$$G_{fi} = (\pi e/\hbar\, m_e c)^2\ |(\tilde{p})_{fi}|^2\ |A_o(\omega_{fi})|^2 \qquad (2\text{-}103')$$

We may express $|A_o(\omega_{fi})|^2$ in terms of the time average of the intensity of the incoming wave.

$$I = \langle \vec{s}\cdot\hat{n}\rangle = (c/4\pi\tau)\int|E|^2 = (c/\tau)\int|E(\omega)|^2\, d\omega \qquad (2\text{-}107)$$

so that

$$I(\omega) \equiv (dI/d\omega) = c|E(\omega)|^2/\tau \qquad (2\text{-}108)$$

where

$$\vec{E}(\omega) = \omega\vec{A}_o(\omega)e^{i\vec{k}\cdot\vec{r}} \qquad (2\text{-}109)$$

Hence

$$|A_o(\omega)|^2/4 = (c^2/\omega^2)|E(\omega)|^2 = (c\tau/\omega^2)I(\omega) \qquad (2\text{-}110)$$

and

$$G_{fi} = (2\pi e/\hbar m_e)^2\ (\tau I(\omega_{fi})/\omega_{fi}^2 c)|(\tilde{p})_{fi}|^2 \qquad (2\text{-}111)$$

The transition probability per unit time is

$$w_a = G_{fi}/\tau = (4\pi^2 e^2/m_e^2\hbar^2 c)(I(\omega_{fi})/\omega_{fi}^2)|(\tilde{p})_{fi}|^2 \qquad (2\text{-}112)$$

<u>Induced Emission</u>. The transition probability per unit time for induced emission of radiation is the same as (2-112) except that ω_{fi}

is replaced by ω_{if} and the integral is replaced by $(e^{-i\vec{k}\cdot\vec{r}}\,\hat{\ell}\,\vec{\nabla})_{fi}$. Interchanging the labels f and i and integrating by parts, using the fact that $\hat{\ell}\cdot\vec{k} = 0$ we find, for induced emission

$$w = (4\pi^2 e^2/\hbar^2 m_e^2 c)(I(\omega)_{fi}/\omega_{fi}^2)\,|(\tilde{p})_{fi}|^2 \qquad (2\text{-}112')$$

which is the same as (2-112) except the sign in the exponential is reversed (see Equation (2-104). Hence the probabilities of reverse transitions between any pair of states under the influence of the same radiation field are the same. This is an example of the principle of detailed balancing. This result was discussed earlier in connection with the Einstein coefficients.

Spontaneous Emission. According to Einstein's theory the probability per second of absorption of a light quantum $h\nu_{fi} = W_f - W_i$ with polarization α and propagated in the solid angle $d\Omega$ is (equation (2-8))

$$dJ_a = (B_{if\alpha} I_\alpha(\nu,\Omega)\,d\Omega/4\pi) \qquad (2\text{-}8')$$

The probability (2-112) was derived on the assumption of a plane wave propagated in a definite direction. For this reason our formula does not involve the angular distribution. In general

$$I_\alpha(\nu) = \int I_\alpha(\nu,\Omega)\,d\Omega \qquad (2\text{-}113)$$

For our case

$$I_\alpha(\nu,\Omega) = I_\alpha(\nu)\delta(\Omega) \tag{2-114}$$

Integrating (2-8′) over solid angles we find

$$w_a = J_a = B_{if\alpha} I_\alpha(\nu_{fi}) = B_{if\alpha} I_\alpha(\omega_{fi})/2 \tag{2-115}$$

$(I(\nu) = 2\pi I(\omega))$ for the absorption probability per second with respect to a wave propagated in a definite direction (with no divergence). Comparing (2-115) and (2-112) the Einstein coefficient for absorption of light is found to be

$$B_{if\alpha} = (2e^2/m_e^2\ \nu_{fi}^2\ \hbar^2 c)|(\tilde{p}_{fi})|^2 \tag{2-116}$$

The coefficient for spontaneous emission is given by the Einstein relation for one polarization is (see Equation (2-20))

$$A_{fi\alpha} = (h\nu_{fi}^3/c^2)B_{if\alpha}$$

$$= (4\pi e^2/\hbar\ m_e^2 c^3)\nu_{fi}|(\tilde{p})_{fi}|^2 \tag{2-117}$$

The probability per second of spontaneous emission of a light quantum $h\nu_{fi} = W_i - W_f$ of polarization α into the solid angle Ω is therefore

$$dJ_s = dw_s = A_{fi\alpha}(d\Omega/4\pi)$$

$$= (e^2\nu_{fi}/\hbar\ m_e^2 c^3)|(\tilde{p}_{fi})|^2 d\Omega \tag{2-118}$$

To include the effects of the spin term

(2-74), the matrix element $(e^{i\vec{k}\circ\vec{r}}\ \hat{\ell}\circ\vec{\nabla})_{fi}$ which appears in equations (2-102), (2-103), (2-111), (2-112) and (2-116) through (2-118) must be replaced by the matrix element

$(\vec{\sigma}\cdot\vec{k}\times\hat{\ell}e^{i\vec{k}\circ\vec{r}})/2$, since $\vec{B} = \vec{\nabla}\times\vec{A} = i\vec{k}\times\vec{A}$.

2.8. The Dipole Approximation

The expressions for the different transition probabilities all contain a term of the form

$$\int u_f^* \, e^{i\vec{k}\circ\vec{r}} \, \hat{\ell}\cdot\vec{\nabla} \, u_i \, dxdydz \qquad (2\text{-}119)$$

To simplify this integral we note that

$$\vec{k}\circ\vec{r} \sim (\nu/c)(h/p) \sim (W/hc)(h/p) \sim W/pc \sim v/c \qquad (2\text{-}120)$$

Hence for non-relativistic problems ($v \ll c$) the exponential can be replaced by unity to first order in v/c. This is just the quantum mechanical analog of the classical dipole approximation (c.f. Section 1.9). Then

$$\int u_f^* \, \hat{\ell}\circ\vec{\nabla} \, u_i \, dxdydz = (i/\hbar)(\hat{\ell}\cdot\vec{p})_{fi} \qquad (2\text{-}121)$$

In the dipole approximation the spin term can be neglected since

$$\frac{H'}{H'_r} = \left|\frac{\vec{\sigma}\cdot\vec{k}\times\hat{\ell}}{\hat{\ell}\circ\vec{\nabla}}\right| \sim \vec{k}\cdot\vec{r} \sim v/c \qquad (2\text{-}122)$$

(see the discussion at the end of the last section).

Using operator algebra, the matrix element

can be written as

$$(\hat{\ell}\cdot\vec{p})_{fi} = (im_e(W_f - W_i)/\hbar)(\hat{\ell}\cdot\vec{r})_{fi} = im_e\omega_{fi}(\hat{\ell}\circ\vec{r})_{fi} \tag{2-123}$$

The transition probability per second for spontaneous emission is then

$$dw_s = 4\pi^2(e^2/\hbar c)(\nu_{fi}^3/c^2)|(\hat{\ell}\circ\vec{r})_{fi}|^2\ d\Omega \tag{2-124}$$

The intensity of the light emitted into the solid angle $d\Omega$ in erg/sec is obtained by multiplying the probability by the energy of the light quantum $W = h\nu$:

$$dP/d\Omega = (8\pi^3 e^2\nu^4/c^3)|(\hat{\ell}\circ\vec{r})_{fi}|^2 \tag{2-125}$$

where we have dropped the subscript fi on ν. If the angle between the direction of observation $\hat{n}$ and the dipole moment $e\vec{r}$ is Θ and the measuring device subtends a solid angle $d\Omega$ at the location of the emitting atom, the observed intensity is

$$dP/d\Omega = (8\pi^3 e^2\nu^4/c^3)\ |(\vec{r})_{fi}|^2\ \sin^2\Theta \tag{2-126}$$

since the direction of polarization is perpendicular to $\hat{n}$. We can without loss of generality, resolve $\hat{\ell}$ into two components, one $\hat{\ell}_1$, perpendicular to $\vec{r}$ and $\hat{n}$ and the other $\hat{\ell}_2$, in a plane determined by $\hat{n}$ and $\vec{r}$ at an angle of $\pi/2 - \Theta$ with $\vec{r}$. Light of polarization $\hat{\ell}_1$ is not emitted at all, and polarization $\hat{\ell}_2$ is emitted with intensity given by (2-126).

The total intensity of emitted radiation is obtained by integrating (2-126) over $d\Omega$:

$$P = (64\pi^4 e^2 \nu^4 / 3c^3) \; |(\vec{r})_{fi}|^2 \qquad (2\text{-}127)$$

The formula is almost identical with the classical formula (1-140) ($\omega = 2\pi\nu$) for radiation from an oscillating dipole described by $\vec{d} = e\vec{r} = d \sin \omega t$. To recover the classical formula we have only to replace the square of the matrix element by the time average of the dipole moment:

$$e^2 (r)^2_{fi} \rightarrow \overline{d^2} = d^2/2$$

This is the usual connection between the classical quantities and quantum theoretical quantities according to the correspondence principle.

The total transition probability for going from i to f is obtained by dividing by $h\nu$.

$$w_{fi} = (32\pi^3/3)(e^2/\hbar c)(\nu^3/c^2) \; |(\vec{r})_{fi}|^2 \qquad (2\text{-}128)$$

If one sums (2-128) over all states f which have energy less than that of the initial state i one arrives at the total probability per unit time that the state i is vacated through radiation

$$w_i = \sum_{f<i} w_{fi} \qquad (2\text{-}129)$$

The reciprocal mean life time of the state i is given by

$$\tau_i = 1/w_i \qquad (2\text{-}130)$$

We can estimate the order of magnitude of

this mean lifetime by using the relations $r \sim \hbar/p$, $p \sim W/v$, $W \sim h\nu$ in equation (2-129):

$$\tau_i \sim (1/\alpha_f)(v/c)^{-2}\ \omega^{-1} \qquad (2\text{-}130')$$

where

$$\alpha_f = e^2/\hbar c = 1/137 \qquad (2\text{-}131)$$

is the <u>fine structure constant</u>, which is in a sense a measurement of the strength of the interaction of the electron with the electromagnetic field. It is often convenient to use the oscillator strength $f(f,i)$:

$$f(f,i) \equiv (4\pi m_e \nu/3\hbar)\ |(\vec{r})_{fi}|^2$$

$$= (h\nu/I_H)|(\vec{r})_{fi}|^2/3a_o^2 \qquad (2\text{-}132)$$

where I_H is the ionization potential of hydrogen and a_o is the Bohr radius:

$$I_H = e^2/2a_o \qquad a_o = \hbar^2/me^2 \qquad (2\text{-}133)$$

Then

$$w(f,i) = (2\pi)\alpha_f^3(I_H/h)(h\nu/I_H)^2\ f(f,i) \qquad (2\text{-}128')$$

$$\simeq 10^9 Z^4\ \sec^{-1} \qquad (2\text{-}128'')$$

where the last estimate is for hydrogenic atoms and ions.

Using the arguments given just below equation (2-97) in Section 2.6, it is easy to show that the total energy (kinetic and potential), W_t, of an electron in the field of a charge Z is a minimum ($= - Z^2 I_H$) for

$$r \sim a_o/z \qquad v \simeq Z\alpha_f c \qquad (2\text{-}134)$$

Therefore $f(f,i) \simeq 1$ and the radiative lifetime is

$$\tau_i \sim 1/Z^2 \alpha_f^3 \omega >> 1/\omega \qquad (2\text{-}130')$$

The radiative lifetime is much longer than the period of oscillation of the dipole, i.e. photon emission is a slow process.

2.9. Density of States

So far we have assumed that the final states of the electron system are well separated, so that the total transition probability w_i given in (2-129) can be evaluated as a simple sum. In practice there are very many closed spaced or even continuously distributed final states. In this case the sum must be replaced by an integral and the expressions for w, A_{fi} and B_{if} must be multiplied by a term which gives the density of final states.

$$w \rightarrow w(dn/dW) \qquad (2\text{-}135)$$

Then the general expression (2-96′) for the transition probability per unit time becomes

$$w = (2\pi/\hbar)(H')^2_{fi}\ (dn/dW) \qquad (2\text{-}136)$$

For many problems the final states are assumed to be that of a free particle, and have wave functions given by $V^{-\frac{1}{2}} \exp(i\vec{k}_f \cdot \vec{r})$ where V is the normalization volume. Thus we can compute the density of final states by analogy with the calculation given in Section 2.2 for number of independent waves in a given volume. There we found that the number of independent waves having a given polarization in the frequency range ν and $\nu+d\nu$ was $(4\pi X^3/c^3)\nu^2 d\nu$, or $(X/c^3)\nu^2\, d\nu d\Omega$ for a solid angle $d\Omega$. Making the substitutions

$$\nu = ck/2\pi;\ k = p/\hbar \tag{2-137}$$

the number of states is

$$dn = (X/h)^3\ p^2\ dp d\Omega \tag{2-138}$$

The density of final states is therefore

$$dn/dW = (dn/dp)(dp/dW) = (X/h)^3 (p^2/v) d\Omega \tag{2-139}$$

since

$$dp/dW = 1/v \tag{2-140}$$

for non-relativistic particles.

2.10. Note on Relativistic Quantum Processes

Throughout this chapter we have assumed that the electrons are non-relativistic. To describe relativistic quantum processes we have to use the relativistic expression for the energy of the electron

$$W^2 = m^2 c^4 + p^2 c^2 \tag{2-141}$$

In addition we have to use the 4-component Dirac wave functions and replace the momentum operator by the operator $mc\vec{\alpha}$, where $\vec{\alpha}$ is given by (c.f. equation (2-70)).

$$\vec{\alpha} = \begin{pmatrix} 0 & \vec{\sigma} \\ \vec{\sigma} & 0 \end{pmatrix} \qquad (2\text{-}142)$$

In principle it is straightforward, but in practice it leads to considerable complications which are beyond the scope of this book. Fortunately a classical description supplemented by the uncertainty principle provides a good idea of the essentials. When detailed formulas are needed, I will merely quote them and give a reference to the literature.

2.11. The Equation of Radiative Transfer; Absorption Coefficient

Assume that radiation of intensity $I(\nu)$ falls on area dA situated on the normal to the direction of radiation incident within a solid angle $d\Omega$, in a frequency interval $(\nu, \nu + d\nu)$, and in time dt. The amount of energy falling on the area is

$$dW_i = I(\nu)\, dA d\Omega d\nu dt \qquad (2\text{-}143)$$

If the medium absorbs the radiation, then some fraction of this energy will be absorbed along the path ds and this fraction will be proportional to ds.

$$dW_a = \mu(\nu) I(\nu) dA d\nu d\Omega dt ds \qquad (2\text{-}144)$$

The amount of energy radiated from the cylindrical volume dAds into $d\Omega$ is

$$dW_r = j(\nu, \Omega) dA d\Omega d\nu dt ds. \qquad (2\text{-}145)$$

In Equation (2-145)

$$j(\nu,\Omega) \equiv N dP(\nu)/d\Omega \quad \text{erg/cm}^3\text{ sec-Hz-ster} \tag{2-146}$$

is the emission coefficient of the gas, and N is the number of emitting atoms electrons per unit volume.

Assuming a radiation field that does not change in time, and equating the energy leaving the cylinder to the energy entering the cylinder plus the energy emitted by the cylinder minus the energy absorbed in the cylinder we obtain an equation which determines the change of intensity of radiation transmitted through an absorbing and emitting medium:

$$dI(\nu)/ds = -\mu(\nu)I(\nu) + j(\nu,\Omega) \tag{2-147}$$

This equation is called the equation of radiative transfer. Casting it into integral form yields

$$I(\nu,s) = I(\nu,o)\exp(-\tau(o,s)) + \\ + \int_o^s j(\nu,\Omega,s')\exp(-\tau(s,s'))ds' \tag{2-148}$$

where

$$\tau(s,s',\nu) = \int_{s'}^{s} \mu(\nu,s')ds' \tag{2-149}$$

is the optical depth for absorption of radiation of frequency ν along the path between s' and s.

This shows that the intensity of radiation consists of two parts. The first part represents the intensity of the original radiation at the point s = 0 diminished by absorption on the path from 0 to s. The second part is an intensity of radiation due to the emission of radiant energy on the path from 0 to s and diminished by the absorption on the way from the place of emission s′ to the given place s.

The absorption coefficient can be defined in terms of the Einstein coefficient A_{mn} or the emission coefficient. The absorption measured by an observer is the difference between the actual absorption in the medium and the induced emission (see Section 2.2). If N_n and N_m are the number of electrons per unit volume in states n and m with energies W_n and W_m such that $W_m - W_n = h\nu$, then the absorption coefficient is

$$\mu(\nu) = -\Delta I/I = (h\nu/4\pi)\sum(N_n B_{nm} - N_m B_{mn})$$

$$= (c^2/8\pi\nu^2)\sum((N_n q_m/q_n) - N_m)A_{mn} \quad (2\text{-}150)$$

When the states are described by a Boltzmann distribution (Equation (2-10)) the absorption coefficient is

$$\mu(\nu) = (c^2/8\pi\nu^2)\sum_m N_m A_{mn}(e^{h\nu/KT} - 1)$$

$$= (c^2/2h\nu^3)(e^{h\nu/KT} - 1)j(\nu,\Omega)) \quad (2\text{-}151)$$

$$= j(\nu,\Omega)/I_{eq}(\nu)$$

where $I_{eq}(\nu)$ is the intensity of black-body radiation (see Equation (2-21)), and the relation (see Equation (2-7))

$$j(\nu,\Omega) = \sum_m N_m A_{mn}/4\pi \tag{2-152}$$

has been used. For continuous states (2-150) is replaced by

$$\mu(\nu) = \int \eta_a(\nu,\vec{p})N(\vec{p})d^3p - \int \eta_i(\nu,\vec{p}')N(\vec{p}')d^3p' \tag{2-153}$$

where η_a and η_i are analogous to the Einstein coefficient B_{nm} and B_{mn}. They give the differential rate absorption and induced emission for electrons in differential momentum interval. $N(\vec{p})$ is the electron distribution function and the momenta $\vec{p}$ and p' are connected by the conservation of momentum and energy in the radiation process ($p^2 > p'^2$). The η_a and η_i are related to the coefficient for spontaneous emission, $dP(\nu,\vec{p})d\Omega$ by relations analogous to (2-12) and (2-20):

$$dP(\nu,\vec{p})/d\Omega = (2h\nu^3/c^2)\eta_i(\nu,\vec{p}')$$
$$\eta_a(\nu,\vec{p})d^3p = \eta_i(\vec{p}')d^3p' \tag{2-154}$$

These relationships can be used to write the absorption coefficient in terms of the differential emission coefficient.

$$\mu(\nu) = (c^2/2h\nu^3)\int (dP(\nu,\vec{p}')/d\Omega)(N(\vec{p}) - N(\vec{p}')) \cdot d^3p' \tag{2-155}$$

For a Maxwell-Boltzmann particle distribution, it follows from the definition

$$j(\nu,\Omega) = \int (dP(\nu,\vec{p})/d\Omega)\ N(\vec{p})\ d^3p \qquad (2\text{-}156)$$

that Equation (2-151) holds.

In the classical limit the energy of the radiated photon is much less than the energy of the electron, so we can expand the integrand of (2-155) about small values of $\vec{p} - \vec{p}'$. If we assume that the distribution function is isotropic then

$$N(p') = N(p) + (\partial N/\partial p)\Delta p$$

$$= N(p) + (h\nu/c)\ (\partial N/\partial p) \qquad (2\text{-}157)$$

and the absorption coefficient is

$$\mu(\nu) = -(c/2\nu^2)\int\ (dp(\nu,\vec{p})/d\Omega)/(\partial N(p)/\partial p)d^3p \qquad (2\text{-}158)$$

The derivative $\partial N(p)/\partial p$ in Equation (2-158) in negative for most distribution functions that apply to astrophysical situations, but there are undoubtedly situations where the derivative is positive over some range of p so that negative absorption, or maser action can occur (see Bekefi, 1966 and Tsytovich, 1973).

The relationships (2-155) and (2-158) refer to unpolarized radiation in a low density medium. For some problems it is necessary to know the absorption coefficient for a given mode of polarization. In this case we must use the Einstein relations for waves with a given polarization $B_{mn} = (c^2/h\nu^3)A_{mn}$ and equations (2-150), (2-151), and (2-155) and (2-158) must be multiplied by a factor of 2. In addition the A_{mn} or the differential emission coefficients refer to an emission of the type of polarized waves under consideration (see Ginzburg and Syrovatskii, 1969).

References

Any good quantum mechanics textbook will serve as a general reference for the material in this chapter. I found the following texts useful:

Bethe and Jackiw (1968)

Blokhintsev (1964)

Houston (1959)

Kramers (1958)

The distinction between antenna temperature and brightness temperature is discussed in Bekefi (1966) and Shklovsky (1960).

The equation of radiative transfer and the absorption coefficient are discussed in Bekefi (1966), Chandrasekhar (1960) and Ginzburg and Syrovatskii (1969).

2.1. Derive Equation (2.5).

2.2. The intensity of radiation from a star is observed to reach a maximum at 6000 Å. The total luminosity of the star is 2×10^{37} ergs/sec. Assuming that the star radiates like a black body, find its temperature and radius.

2.3. The cosmic microwave background radiation field fits a black body curve with a temperature of $2.7^{\circ}K$. What's the mean energy per photon? What's the energy density of the radiation? How does it compare with the energy density in our galaxy of starlight, cosmic rays, and magnetic fields?

2.4. Derive the relationship between antenna temperature and brightness temperature for a source whose angular dimension is small compared to the effective beam width.

2.5. Consider the wave function

$$\psi(x) = (a/\pi)^{\frac{1}{4}} e^{-ax^2/2},$$

which represents the ground state of a harmonic oscillator. Evaluate the average value of x, x^2, p, and p^2, and show that the product

$$\overline{(x - \overline{x})^2}\,\overline{(p - \overline{p})^2} = \hbar^2/4$$

Evaluate this quantity for any other properly normalized wave function.

2.6. Verify Equation (2-56).

2.7. Show that the spin functions defined by Equation (2-73) are eigenfunctions of S^2 and S_z.

2.8. Using Equation (2-10), derive the Saha ionization formula

$$N_1 N_e/N_0 = (u_1/u_0) 2(2\pi m_e KT)^{3/2} h^{-3} e^{-I_0/KT}$$

where N_1 is the number density of singly ionized ions, N_0 is the number density of neutrals, and N_e is the number density of electrons. I_0 is the ionization energy of the neutral atom, and the u's are the partition functions of the atom and ion:

$$u_0 = \sum q_{0,n} e^{-W_{0,n}/KT}, \text{ etc.}$$

The statistical weight of the ionized state is the product of the statistical weights of the ion and the free electron. The statistical weight of the electron can be computed from Equations (2-138) and (5-57).

Chapter 3

CYCLOTRON AND SYNCHROTRON RADIATION

Charged particles moving in a magnetic field experience acceleration as long as their motion is not parallel to the field, and therefore, according to classical electrodynamics they must emit electromagnetic waves. This emission is called cyclotron radiation if the electrons are non-relativistic or synchrotron radiation if the electrons are relativistic. In work by Russian authors, the radiation is often called magnetobremsstrahlung (Ginzburg and Syrovatskii 1965, 1969). The radiation from charged particles moving in a magnetic field was first discussed many years ago by Schott (1912). Interest in synchrotron radiation was revived in 1945 and the following years in connection with problems such as the radiation from electron accelerators, charged particles in the earth's magnetic field, and cosmic radio sources.

In astrophysics, an understanding of synchrotron radiation is especially important, since non-thermal cosmic radio emission in a majority of cases appears to be due to this process. This is true for general galactic radio emission, radio emission from supernova remnants, and radio emission from radio galaxies. Synchrotron radiation is also observed in sporadic radio emission from the Sun and Jupiter. In addition optical and x-ray synchrotron radiation is observed from the Crab Nebula, many radio galaxies, and possibly quasi-stellar sources.

Cyclotron radiation is in general less important. However, it does play an important role in certain high magnetic field situations such as found in solar flares (radio), white dwarfs (optical), and neutron stars (X-ray).

The first part of this chapter is devoted to a discussion of cyclotron and synchrotron radiation from individual particles and ensembles of particles. Many of the basic formulas have been derived in Chapter 1, so not much space is given to either heuristic or formal derivations. For these, the reader may refer back to Chapter 1 or to the reviews by Ginzburg and Syrovatskii (1965), Blumenthal and Gould (1970 and Pacholczyk (1970). In the second part astrophysical applications of the synchrotron formulas are discussed. More applications are given in the problems at the end of this chapter.

3.1. Total Power Emitted, Peak Frequency and Radiative Lifetime

Consider an electron with energy W-$\gamma m_e c^2$ in a uniform magnetic field B with components of velocity $c\beta_{\parallel}$ and $c\beta_{\perp}$ parallel and perpendicular to the field. Ignoring the force of radiative reaction, the equation of motion is given by

$$d(\gamma m_e c\vec{\beta})/dt = e\ \vec{\beta} \times \vec{B}, \qquad (3\text{-}1)$$

and the particle experiences an acceleration perpendicular to its velocity. Therefore, γ, β, and $\beta_{\parallel}$ are constants of the motion as is the pitch angle α between $\vec{\beta}$ and $\vec{B}$ since $\cos\alpha = \beta_{\parallel}/\beta$. Therefore, from Equation (3.1) the perpendicular component of velocity obeys the equation

$$d\vec{\beta}_{\perp}/dt = \vec{\beta}_{\perp} \times \vec{\omega}_o \qquad (3\text{-}2)$$

where $\beta_{\perp} = \beta \sin\alpha$ and

$$\vec{\omega}_o = \vec{\omega}_B/\gamma = (e\vec{B}/\gamma m_e c) = 1.8\times10^7\ \vec{B}/\gamma \quad \text{rad/sec} \qquad (3\text{-}3)$$

The path of the particle is a circular helix of radius

$$R_B = m_e\gamma\ \beta_{\perp} c^2/eB \qquad (3\text{-}4)$$

and pitch angle α.

In Section 1.10 the total power radiated by a charged particle of charge e and mass m_e moving in a magnetic field was found to be

$$dW/dt = P = (2r_o^2/3c)\gamma^2\beta^2 B^2\ \sin^2\alpha \qquad (3\text{-}5)$$

where $r_o = e^2/m_e c^2$ is the <u>classical electron radius</u>.

Since for a given particle velocity the power radiated is proportional to m^{-2}, radiation from electrons is far more important than from protons. For electrons the numerical value for the power is

$$P = 1.6\times10^{-15}\ \gamma^2 B^2\beta^2\ \sin^2\alpha \qquad \text{erg/sec} \quad (3\text{-}5')$$

For a nonrelativistic electron the power is proportional to the energy of the electron, whereas for a relativistic electron it is proportional to the square of the energy. For an isotropic distribution of electron velocities

$$P = (4/9)r_o^2 c\ \gamma^2 B^2\beta^2$$

$$= 1.1\times10^{-15}\ \gamma^2 B^2 \beta^2 \quad \text{erg/sec} \qquad (3\text{-}6)$$

In the nonrelativistic case most of the power is radiated at the frequency $\nu = \omega_o/2\pi$, with ω_o given by equation (3.3). In the relativistic case, most of the power is radiated near the peak frequency

$$\nu_m \simeq c\gamma^3/2\pi R_B = \nu_B\gamma^2 \sin\alpha = 3\times10^6 B\gamma^2 \sin\alpha \text{ Hz}$$

$$= 1\times10^7 \ B\ W^2_{MeV} \sin\alpha \qquad \text{Hz} \qquad (3\text{-}7)$$

where W_{MeV} is the energy of the electron in units of MeV.

The time for electrons to lose a significant fraction of their energy by radiation is

$$t_{sy} = W/P = (3\times10^8/\gamma B^2 \beta^2 \sin^2\alpha) \qquad \text{sec} \qquad (3\text{-}8)$$

This can be rewritten in terms of the average radiated frequency by using Equation (3-7):

$$t_{sy} = (5\times10^{11}/B^{3/2}\ \nu_m^{1/2}\ \sin^{3/2}\alpha) \qquad \text{sec} \qquad (3\text{-}9)$$

The lifetime decreases with increasing frequency radiated and increasing magnetic field. For some sources t_{sy} is small compared to the lifetime of the source, which implies that continuous injection of energetic electrons is necessary to maintain the source. For example, in the case of the Crab Nebula, the lifetime of the electrons producing the 10^{18} Hz X-rays in a magnetic field of the order of 10^{-4} gauss is about 20 years, whereas the Crab Nebula is known to be over 900 years old. This argument led to the conclusion that acceleration of electrons up to energies of 10^8

MeV must still be occurring.

Just how this was accomplished was a great mystery which was solved by the discovery of a pulsar in the center of the Crab Nebula (Hewish, 1970, Ruderman, 1972 and references cited therein).

The fractional energy radiated per cycle is

$$(2\pi/\omega_o t_{sy}) = 10^{-15}\ B\gamma^2 \qquad (3\text{-}10)$$

Equations (3-7) and (3-10) show that the fractional energy radiated per cycle is negligible for peak frequencies much less than 2×10^{21} Hz, corresponding to photon energies much less than 10 MeV. Therefore, except for the gamma-ray region, the force of radiative reaction can be neglected and the electron energy can be assumed to be constant when computing its trajectory over a few orbits.

Quantum mechanical effects are negligible so long as the energy of the photons are much less than the kinetic energy of the particles. In the relativistic case this means that B and γ must satisfy the condition $B\gamma^2 \ll 5\times10^{13}$, or, in terms of the peak frequency,

$$B \ll 10^{17}/\nu_m^{1/2} \quad \text{gauss.} \qquad (3\text{-}11)$$

Quantum effects are important near the surface of neutron stars, where B is of the order of 10^{12} gauss, but are unimportant elsewhere. For the non-relativistic case, the condition for the neglect of quantum effects becomes

$$B \ll 10^8\ W_{eV} \quad \text{gauss} \qquad (3\text{-}12)$$

Again, we see that quantum effects are unimportant, except for neutron stars and perhaps high magnetic field white dwarfs.

3.2. Spectrum of Radiation From An Electron Moving Perpendicular To The Magnetic Field

In this case the electron's velocity is given by

$$\vec{\beta} = \beta(\hat{x} \cos\omega_o t + \hat{y} \sin\omega_o t) \tag{3-13}$$

and the position vector

$$\vec{R}' = \vec{R}_B(\hat{x} \sin\omega_o t - \hat{y} \cos\omega_o t) \tag{3-14}$$

(magnetic field in z direction).

The basic character of the radiation is easily understood. In the non-relativistic case, the dipole approximation applies (Section 1.9) and most of the radiation is emitted almost isotropically at the frequency of rotation,

$$\nu_B = eB/2\pi m_e c \tag{3-15}$$

(Equations (1-149), (3-3), (Fig. 1.2)). In the relativistic case the Doppler effect becomes important. The radiation is beamed into a narrow cone about the direction the observed radiation (see Fig. 1.3), and the spectrum is quasi-continuous, consisting of closely spaced lines of frequency on the order of the peak frequency given by Equation (3-7), and separated by $\nu_o = \omega_o/2\pi$. The total power emitted is much greater for relativistic motion than for non-relativistic motion (Equation 3-5).

The spectral distribution of the radiation is obtained by substituting this expression into equation (1-183″):

$$dP_s/d\Omega = (e^2\omega^2/8\pi^3 c)\left|\int_0^\tau \hat{n}\times(\hat{n}\times\vec{\beta})\exp i\omega(t-n\cdot R'/c) \circ dt'\right|^2 \tag{3-16}$$

where

$$\omega = s\omega_0 \qquad (s = 1,2,3,\ldots)$$

$$\tau = 2\pi/\omega_0 \tag{3-17}$$

To evaluate the integral (3-16) for arbitrary β we have assumed that

$$\hat{n} = \hat{x}\sin a + \hat{z}\cos a \tag{3-18}$$

Then the integral (3-16) is reduced to the problem of evaluating integrals of the form (see Landau and Lifshitz, 1961)

$$\int_0^{2\pi} \cos u\, e^{is(u-\beta\sin a\sin u)}\,du = (2\pi/\beta\sin a)J_s(s\beta\sin a) \tag{3-19}$$

$$\int_0^{2\pi} \sin u\, e^{is(u-\beta\sin a\sin u)}\,du = 2\pi i\, J'_s(s\,\beta\,\sin a) \tag{3-20}$$

where the $J_s(x)$ is the Bessel function of integral order s, and the prime on the Bessel function denotes differentiation with respect to its argument. The intensity of the radiation is therefore

$$dP_s/d\Omega = (2\pi e^2 s^2 \nu_0^2/c)[\cot^2 a J_s^2(s\beta \sin a) + \beta^2 J_s'^2(s\beta \sin a)] \quad (3\text{-}21)$$

The integration over angles can be performed by using Bessel function identities. The result is

$$P_s = (8\pi^2 e^2 \nu_0^2/v)\{s\,\beta^2 J'_{2s}(2s\beta) - s^2\gamma^{-2}\int_0^\beta J_{2s}(2su)du\} \quad (3\text{-}22)$$

Summing over all harmonics, we recover expression (3-5) for the total power radiated.

3.3. Cyclotron Radiation From Non-Relativistic Electrons

When $s\beta$ is much less than unity, the Bessel functions can be replaced by their asymptotic series expansions for small arguments:

$$J_n(z) = z^n/2^n n! \quad (3\text{-}23)$$

The dominant term in P_s is

$$P_s = (8\pi^2 \nu_B^2 e^2/c)(s+1)s^{2s+1}\,\beta^{2s}/(2s+1)! \quad (3\text{-}24)$$

($\omega_o = \omega_B$ for non-relativistic motion).

The ratio of power in successive harmonics is

$$P_{s+1}/P_s = \beta^2 \qquad (3\text{-}25)$$

so almost all the energy is radiated in the first harmonic.

The angular distribution of the radiation in the first harmonic is almost isotropic:

$$dP_1/d\Omega = (\pi e^2 \nu_B^2 \beta^2/2c)(1 + \cos^2 a) \qquad (3\text{-}26)$$

This expression could also have been derived from Equation (1-149) and (3-14).

The width of the cyclotron lines is determined by a number of broadening mechanisms. These mechanisms are discussed in detail by Bekefi (1966). They are:

(a) Radiation broadening: $\Delta\nu \simeq 1/t_{sy}$ with t_{sy} given by (3-8).

(b) Collisional broadening: $\Delta\nu \simeq$ collision frequency.

(c) Doppler broadening: $\Delta\nu \simeq (v_{th}/c)\nu$ where v_{th} is the thermal velocity of the radiating charge.

(d) Self-Absorption of Radiation: When the optical thickness at the line center exceeds unity the radiation intensity approaches that of a black body. In a thermal plasma the intensity cannot exceed the black body intensity so the radiation is distributed over a wider frequency range. This is often the dominant broadening mechanism in astrophysical situations.

(e) Plasma dispersion effects: Near the plasma frequency (Equation (1-186)) the dispersive properties of the plasma affect the line shape and width.

(f) Non-uniformity in the magnetic field: $\Delta\nu \simeq (e/2\pi\ m_e c)\Delta B$.

3.4. Synchrotron Radiation From Ultrarelativistic Electrons

For ultrarelativistic electrons β is approximately unity, γ is large and

$$1 - \beta \simeq 1/2\gamma^2 \tag{3-27}$$

From the discussion in Section 1.10 and references given therein we know that most of the radiation in the ultrarelativistic case is emitted into a narrow cone with a half-angle $\theta \sim 1/\gamma$ about the direction of motion of the electron. The electron radiates in a given direction only for a short time $\Delta t \sim 1/\nu_0\gamma$. For times much greater than this the exponential term in equation (3-16) or (3-20) oscillates rapidly and the integrals averages out to zero. Therefore most of the radiation occurs at frequencies $\gg \nu_0$, i.e., at high harmonics where $s \gg 1$. Since most of the radiation is emitted in the plane of motion which is perpendicular to the magnetic field the angles $a \simeq \pi/2$ and $\sin a \simeq 1$. Since most of the radiation is emitted at high harmonics where the spectrum is practically continuous it is more convenient to discuss the power radiated per unit frequency interval. From Equation (3-17) we have

$$dP(\nu)/d\Omega = (dP_s/d\Omega)(ds/d\nu) = (1/\nu_0)(dP_s/d\Omega)$$

(3-28)

In evaluating the expression (3-21) asymptotic expressions for Bessel functions with large argument and order can be used, together with the relation

$$1 - \beta^2 \cos^2\theta \approx \theta^2 + \gamma^{-2} \qquad (3\text{-}29)$$

The intensity of the synchrotron radiation is therefore

$$dP(\nu)/d\Omega = (2e^2\nu^2/3\pi c\nu_B\gamma^3)[\theta^2\gamma^2(1+\theta^2\gamma^2) \circ K^2_{1/3}(\xi) + (1+\theta^2\gamma^2)^2 K^2_{2/3}(\xi)] \qquad (3\text{-}30)$$

where

$$\xi = (1/3)(\nu/\nu_0)(\theta^2 + \gamma^{-2})^{3/2} \qquad (3\text{-}31)$$

The first term in the brackets corresponds to radiation polarized perpendicular to the plane of the orbit, and the second to radiation polarized in that plane.

To examine this result further, we make use of the limiting forms of the modified Bessel functions:

$$K_n(x) \simeq \Gamma(n)\ (2/x)^n/2 \qquad n \neq 0\ ,\ x \ll 1$$

$$K_n(x) \simeq (\pi/2x)^{\frac{1}{2}}\ e^{-x} \qquad x \gg 1 \qquad (3\text{-}32)$$

Thus for $\xi \ll 1$, the intensity of the radiation is a weak function of ξ; for $\xi \gg 1$ the intensity of the radiation will be negligible. Equation (3-31) shows that $\xi \gg 1$ at large angles. As the frequency increases, the critical angle beyond which there will be negligible radiation decreases. If ν gets too large ξ will be large at all angles and there will be a negligible amount of energy radiated. The critical frequency beyond which there is negligible radiation at any angle can be defined by $\xi = 1$ for $\theta = 0$:

$$\nu_c = (3/2)\gamma^2 \nu_B = 4\times10^6 \ B \ \gamma^2 \tag{3-33}$$

Integration of (3-30) over angles yields the result:

$$P(\nu) = (2\pi \sqrt{3}\ e^2 \nu_B/c)\ F(\nu/\nu_c) \quad \text{erg/sec-Hz} \tag{3-34}$$

where

$$F(x) = x \int_x^\infty K_{5/3}(y)dy \tag{3-35}$$

and $K_{5/3}$ is the modified Bessel function of order 5/3. F(x) or, equivalently, the non-dimensionalized synchrotron spectrum is plotted in Figure 3.1. This function reaches its maximum value of 0.918 at x = 0.29. Well away from the maximum F(x) asymptotically approaches the limits

$$F(x) \to \begin{cases} (4\pi/\sqrt{3}\ \Gamma(1/3))(x/2)^{1/3} & \text{for } x \ll 1 \\ (\pi/2)^{1/2}\ x^{1/2}\ e^{-x} & \text{for } x \gg 1 \end{cases} \tag{3-36}$$

The slow decrease ($\propto\nu^{1/3}$) toward low frequency

of the synchrotron spectrum is an important feature. Since the peak frequency and the power emitted vary as γ^2, in the absence of absorption, small pitch angles and plasma effects, a monoenergetic distribution of particles will give the most sharply peaked distribution possible. Thus, if a cut-off sharper than $\nu^{1/3}$ is observed, we must assume that one of these effects is operating.

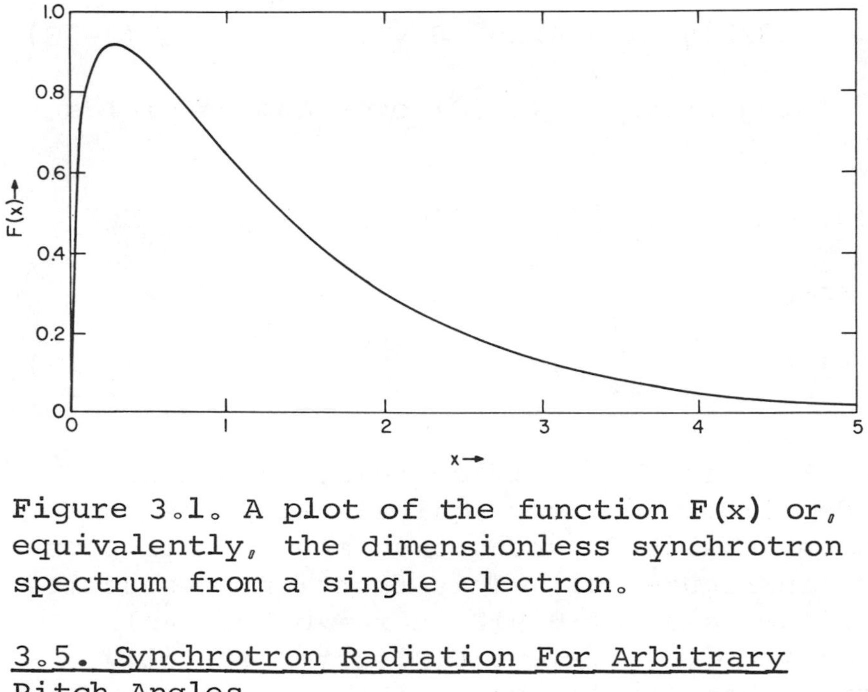

Figure 3.1. A plot of the function F(x) or, equivalently, the dimensionless synchrotron spectrum from a single electron.

3.5. Synchrotron Radiation For Arbitrary Pitch Angles

Equation (3-34) for the emitted spectral power is valid only for pitch angle $\alpha = \pi/2$. To calculate the synchrotron distribution in the laboratory system when $\beta_{\parallel} \neq 0$, we transform to a coordinate system moving with velocity $\beta_{\parallel}$ with respect to the laboratory frame, compute the radiation according to pitch angle

$\alpha' = \pi/2$ and equation (3-34), and transform back to the lab frame. The result is (see Blumenthal and Gould 1970).

$$P(\nu,\alpha) = (\sqrt{3}\, e^3 B \sin\alpha / mc^2) F(\nu/\nu_c)$$

$$= 2.3\times10^{-22}\, B \sin\alpha\, F(\nu/\nu_c) \qquad (3\text{-}37)$$

where the critical synchrotron frequency is given by

$$\nu_c = (3\nu_B/2)\gamma^2 \sin\alpha = 4.3\times10^6 B\, \gamma^2 \sin\alpha \qquad (3\text{-}38)$$

Thus, the spectral shape of synchrotron radiation is independent of pitch angle.

When a particle emits synchrotron radiation while spiraling in a magnetic field with pitch angle α, the radiation is essentially all emitted within an angle $\alpha + O(1/\gamma)$ to the magnetic field. Therefore, when an observer sees radiation from a single particle, that particle has a net (averaged over period) velocity in the direction of the observer of $c\beta_{\|}\cos\alpha = c\beta\cos^2\alpha$. The time between successive pulses as seen by a distant observer must be shorter than $\tau = 2\pi/\omega_0$, the period of emission, by an amount $\beta\cos^2\alpha$ as illustrated in Figure 3.2. Thus, for $\gamma \gg 1$, $\tau_{obs} = \tau\sin^2\alpha$. However, because of conservation of energy, the total energy emitted per period must equal the energy observed in a period, and

$$P_{obs} = P\tau/\tau_{obs} = P/\sin^2\alpha \qquad (3\text{-}39)$$

Both the total power and spectral power

observed are larger than the emitted power given in Equations (3-5) and (3-37) by a factor of $1/\sin^2\alpha$. This result is a consequence of energy conservation, where the total emitted power equals the energy flux through a surface at the observer plus the change in total energy contained within the surface, this change being due to the decreasing average distance between the electron and observer. For a steady distribution of particles confined within a fixed region, the emitted and received powers are identical since the average distance between the particles and the observer does not change with time and it is therefore necessary to calculate only the total emitted synchrotron power.

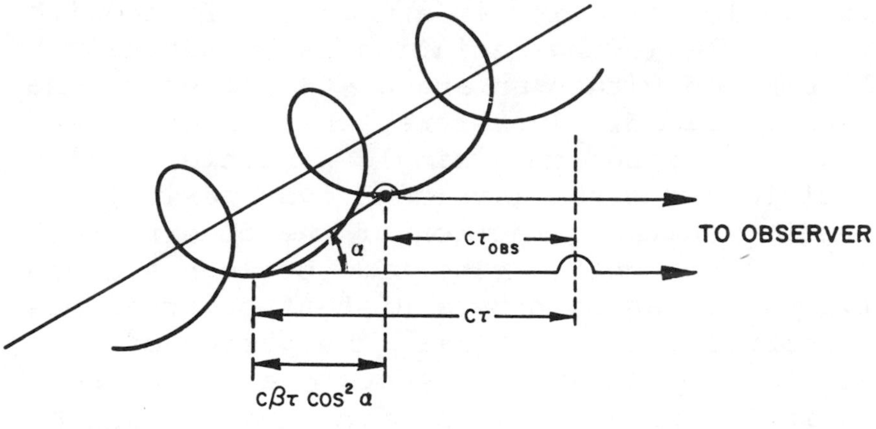

Figure 3.2. A picture of an electron spiraling in a magnetic field showing the difference between the period of emission τ and the observed period τ_{obs}. The dot represents the electron as it emits a pulse of radiation towards the observer.

Equation (3-37) was derived on the assumption that $\alpha \gg \gamma^{-1}$. This is valid except for a very small range of solid angles, and therefore yields a good approximation when one considers a fairly isotropic distribution of electrons. However, situations can occur where the particles spiral with very small pitch angles $\alpha \lesssim \gamma^{-1}$ with the magnetic field. This can occur when the particles are preferentially accelerated along $\vec{B}$, or when energy losses have depleted the particles with large pitch angles ($P \propto \sin^2\alpha$). In the case where $\alpha \ll \gamma^{-1}$, the electron is non-relativistic in the coordinate frame where $\beta_{\parallel} = 0$. Then, in that frame the emission is cyclotron emission, which is monochromatic at the cyclotron frequency and is emitted with the dipole pattern characteristic of non-relativistic emission (Equation 3-26). Using the same technique which yielded Equation (3-37), one obtains

$$P(\nu) = \begin{cases} (2\pi^2 e^2 \alpha^2 \nu/c)(1 - (\nu/\gamma\nu_B) \\ \quad + (1/2)(\nu/\gamma\nu_B)^2); & \nu_B/\gamma \ll \nu \leq 2\gamma\nu_B \\ 0 \quad ; & \nu > 2\gamma\nu_B \end{cases} \tag{3-40}$$

Note that the peak frequency in this case is $2\gamma\nu_B$, a factor of γ^{-1} smaller than the large pitch angle critical frequency (Equation (3-38)).

3.6. Synchrotron Radiation From An Ensemble of Particles

When the density of relativistic electrons at a point with Lorentz factor between γ and

$\gamma + d\gamma$ and pitch angle between α and $\alpha + d\alpha$ is given by $N(\gamma,\alpha,t)d\gamma d\Omega_\alpha$, where $d\Omega_\alpha = 2\pi \sin \alpha d\alpha$, then the spectral emission per unit volume can be calculated using the single particle synchrotron spectrum. If $j(\nu)$ is the total spectral power emitted per unit volume and if the radiation from the individual particles is incoherent, then

$$j(\nu) = \int N(\gamma,\alpha,t)P(\nu)d\gamma d\Omega_\alpha \tag{3-41}$$

When the distribution function is not time dependent the emitted and received synchrotron spectra from a fixed volume are equal. For a time dependent distribution function, t must be replaced by t-R/c, where R is the distance of the particles from the observer.

For many cases of interest in astrophysics the distribution of relativistic electrons can be approximated by a power law:

$$N(\gamma,\alpha) = N_r \gamma^{-n}\, g(\alpha)/4\pi \tag{3-42}$$

for γ in some range $\gamma_1 < \gamma < \gamma_2$. For an isotropic distribution $g(\alpha) = 1$. If we substitute (3-42) and (3-37) into (3-41) we obtain

$$j(\nu) = N_r (3^{\frac{1}{2}}e^2/2c)\nu_B(2\nu/3\nu_B)^{(1-n)/2}$$

$$\cdot\, H(\alpha)(G(x_2) - G(x_1)) \tag{3-43}$$

where

$$x = \nu/\nu_c \tag{3-44}$$

$$H(\alpha) = \int (\sin\alpha)^{(n+1)/2}\, g(\alpha)d\Omega_\alpha \tag{3-45}$$

and

$$G(x_1) = \int_{x_1}^{\infty} x^{(n-1)/2} \int_{x}^{\infty} K_{5/3}(h)\,dy\,dx \qquad (3\text{-}46)$$

G(x) has the following limiting properties:

$$G(\infty) = 0\ ;\ G(0) = 2^{(n-3)/2}\ \frac{3n+7}{3(n+1)}$$

$$\cdot\ \Gamma\left(\frac{3n-1}{12}\right)\Gamma\left(\frac{3n+7}{12}\right) \qquad (3\text{-}47)$$

($\Gamma(x)$ is the gamma function).
When $x_2 \ll 1 \ll x_1$, $G(x_2) - G(x_1)$ is independent of the frequency and

$$j(\nu) \propto \nu^{(1-n)/2} \qquad (3\text{-}48)$$

independent of the pitch angle distribution. Then the indices characterizing the synchrotron and the electron spectra are related by

$$n = 2q + 1 \qquad (3\text{-}49)$$

where q is the synchrotron spectral index

Note that when $\nu \gg \nu_B \gamma_2^2$ the synchrotron spectrum should fall off exponentially (see Equation (3-36)).

In the case of local isotropy of the electron distribution $g(\alpha) = 1$; integrating (3-43) over α yields the result

$$j(\nu) = (8\pi^2 N_r e^2/c)\nu_B a(n)(3\nu_B/2\nu)^q$$

$$= 1.7\times10^{-21}\ N_r a(n) B^{q+1}(4\times10^6/\nu)^q \qquad (3\text{-}50)$$

where

$$a(n) = (2^q \sqrt{3}/8\pi^{\frac{1}{2}})\Gamma(\frac{3n-1}{12})\Gamma(\frac{3n+19}{12})\Gamma(\frac{n+5}{4})$$

$$\cdot\ ((n+1)\Gamma(n+7/4))^{-\frac{1}{2}} \qquad (3\text{-}51)$$

The function a(n) has been tabulated by Ginzburg and Syrovatskii (1965). Their results are reproduced in Table 3.1.

In synchrotron sources, energy losses due to radiation can modify the electron's distribution function so that it is roughly given by a power law with a sharp cut off. Then for an initially isotropic distribution of pitch angles the resulting spectrum will have a bend or knee at a frequency which varies with time (see Section 3.9).

In calculating the synchrotron spectrum from a power law distribution of electrons, Equation (3-43), it was assumed that in the coordinate frame where $\beta_{\parallel} = 0$, the particle's motion is relativistic, or that the pitch angle $\alpha >> \gamma^{-1}$. It was also assumed that the frequency was between the critical frequencies associated with the endpoints of the electron distribution. When either of these assumptions is not valid, the result must be modified. When a power law distribution of electrons, emits synchrotron radiation with $\alpha\gamma << 1$, then for $\nu_B\gamma_1 << \nu << \nu_B\gamma_2$ the spectral index of the resulting emission is 2-n rather than the usual synchrotron result of (1-n)/2 (see Epstein, 1973).

Even when $\gamma\alpha >> 1$, the usual synchrotron result is no longer valid for frequencies below $\nu_B\gamma_1^2$. O'Dell and Sartori (1970) have derived an approximation for the spectrum in this case. They ignore the discrete

Table 3.1

Synchrotron Functions Defined In Equations (3-51) and (3-59)*

n	1	1.5	2	2.5	3	5
a(n)	0.283	0.147	0.103	0.0852	0.0742	0.0922
$y_1(n)$	0.80	1.3	1.8	2.2	2.7	4.0
$y_2(n)$	0.00045	0.911	0.032	0.10	0.18	0.65

* from Ginzburg and Syrovatskii (1965)

nature of the low-frequency single particle synchrotron emission and assume that $P(\nu) = 0$ for $\nu < \nu_o/\sin^2\alpha$ and that $P(\nu) \propto (\nu \sin^2\alpha)^{1/3}$ for $\nu_o/\sin^2\alpha < \nu << \nu_c$. Note that the observed gyration frequency is given by $\nu_o/\sin^2\alpha$ since $\tau_{obs} = \tau \sin^2\alpha$ (see Section 3.5). Since

$\nu_o \propto \gamma^{-1}$ while $\nu_c \propto \gamma^2$, then for a given pitch angle both the lowest and highest frequency is emitted by the highest energy particles. Thus, for an electron distribution $N(\gamma) \propto \gamma^{-n}$, the resulting spectral power for a fixed pitch angle is $dP(\nu)/dV \propto \nu^n (\sin\alpha)^{2n}$ for $\nu << \nu_B/\sin\alpha$, while the standard result given by equation (3-43) holds for $\nu >> \nu_B/\sin\alpha$. This so-called cyclotron turnover between the ν^n low frequency spectrum and the $\nu^{(1-n)/2}$ high frequency spectrum occurs at a frequency

$$\nu_{ct} \sim \nu_B/\sin\alpha \tag{3-52}$$

3.7. Polarization

Synchrotron radiation in general exhibits elliptical polarization. To obtain an expression for the linear polarization, one must decompose the electric vector into two components: $\hat{\ell}_2$ perpendicular to both $\vec{B}$ and the direction of propagation and $\hat{\ell}_1$ perpendicular to both $\hat{\ell}_2$ and the direction of propagation. Then if $P_j(\nu)$ is the spectral power radiated in the jth component, the linear polarization is defined as

$$\Pi = \frac{P_2(\nu) - P_1(\nu)}{P(\nu)} \tag{3-53}$$

The result for a single particle spiraling in a magnetic field is

$$\Pi = x\, K_{2/3}(x)/F(x) \qquad (3\text{-}54)$$

where $x = \nu/\nu_c$ and $K_{2/3}(x)$ is the modified Bessel function of order 2/3. The polarization goes from 0.5 for $\nu/\nu_c \ll 1$ to unity at $\nu/\nu_c \gg 1$.

To obtain the degree of linear polarization from a distribution of radiating electrons, one can evaluate the two integrals corresponding to Equation (3-41). The expression for $P_1(\nu)$ and $P_2(\nu)$ can be obtained from Equation (3-54) and

$$P(\nu) = P_1(\nu) + P_2(\nu) \qquad (3\text{-}55)$$

For the power law electron spectrum given in Equation (3-42) the linear polarization is given by

$$\Pi = (n+1)/(n+(7/3)) \qquad (3\text{-}56)$$

for $\nu_c(\gamma_1) \ll \nu \ll \nu_c(\gamma_2)$, and is independent of frequency in this range. The polarizations observed from cosmic radio sources are usually smaller than this due to the effect of non-uniform magnetic fields and the Faraday rotation of the propagating radiation. There is some small circular polarization associated with synchrotron emission. For a power law electron spectrum the degree of circular polarization is roughly $\sim(\nu_B/\nu)^{1/2} \sim \gamma^{-1} \ll 1$. (Legg and Westfold, 1968).

3.8. The Energy Content Of The Relativistic Electrons In Cosmic Radio Sources

The most important application of the synchrotron theory has been its use to estimate the energy content of the relativistic electron in cosmic radio sources. In many cases the radiation spectrum of a radio source can be represented with sufficient accuracy by a power law: $F(\nu) \propto \nu^{-q}$. Assuming that the radiation is due to the synchrotron process, we immediately obtain the exponent n of the differential electron energy spectrum from the relation $n = 2q + 1$ (Equation (3-49)). If the magnetic field in the radiating region is assumed to be on the average directed at random over the line of sight, and the number density of particles is assumed to be uniform, throughout the volume V of the source, then we can use (3-50) to express the normalizing constant N of the electron spectrum in terms of B, V, and $j(\nu)$. From this we can determine the total number of relativistic electrons $\mathcal{N}_r$ and the total energy of electrons $\mathcal{U}_r$ in the source responsible for the radiation in the observed frequency interval $\nu_1 \leq \nu \leq \nu_2$. We have

$$\mathcal{N}_r = N_r V \int_{\gamma_1}^{\gamma_2} \gamma^{-n} \, d\gamma \tag{3-57}$$

$$\mathcal{U}_r = m_e c^2 \; N_r V \int_{\gamma_1}^{\gamma_2} \gamma^{(1-n)} \, d\gamma \tag{3-58}$$

Here γ_1 and γ_2 are the limits of the energy interval within which the electron spectrum has the form $N_r \gamma^{-n}$. The frequencies ν_1 and

ν_2 are connected with γ_1 and γ_2 by the following approximate relations (see Ginzburg and Syrovatskii, 1965).

$$\gamma_1 = (2\nu_1 / 3\nu_B y_1(n))^{\frac{1}{2}}$$
$$\gamma_2 = (2\nu_2 / 3\nu_B y_2(n))^{\frac{1}{2}} \tag{3-59}$$

where $y_1(n)$ and $y_2(n)$ are numerical factors tabulated in Table 3.1. In most cases $\nu_1 >> \nu_2$, so, eliminating $N_r V$ through the expressions (3-50) and

$$L(\nu) = \int j(\nu)\, dV \quad \text{erg/sec-Hz} \tag{3-60}$$

we find

$$\eta_r = 6\times10^{20}\ L(\nu_1)B^{-1}\ y_1(n)^{(n-1)/2}/(n-1)a(n) \tag{3-61}$$

$$\mathcal{U}_r = \frac{2\times10^{11} B^{-3/2}}{a(n)} \times \begin{cases} L(\nu_1)\nu_1^{\frac{1}{2}}\ y_1(n)^{(n-2)/2}(n-2) & n > 2 \\ L(\nu_2)\nu_2^{\frac{1}{2}}\log(\nu_2 y_1(2)/\nu_1 y_2(2)) & n = 2 \\ L(\nu_2)\nu_2^{\frac{1}{2}}\ y_2(n)^{2-n/2}(2-n) & 1/3 < n < 2 \end{cases} \tag{3-62}$$

For most cases it is sufficient to use the

simple estimate (see Equation (3-9))

$$\mathcal{U}_r \cong L\, t_{sy} = 5\times10^{11}\ B^{-3/2}\ L\nu_*^{-\frac{1}{2}}\ \text{ergs} \qquad (3\text{-}63)$$

which is accurate to within a factor 3 if we choose $\nu_* = \nu_1$ for $n > 2$, and $\nu_* = \nu_2$ for $n \leq 2$. L is the total luminosity in the range ν_1 to ν_2

$$L = \int_1^2 L(\nu)d\nu \qquad \text{erg/sec} \qquad (3\text{-}64)$$

In the near future observations of the X-rays produced by Compton scattering of the microwave background radiation by the relativistic electrons which produce the radio emission should enable us to estimate the magnetic field strength and therefore the energy radio sources (see Chapter 4). A common assumption is that the energies in the magnetic field and relativistic particles in the source are approximately equal.

$$\mathcal{U}_{mag} = \int B^2\ dV/8\pi = q_m \mathcal{U}_r \qquad (3\text{-}65)$$

where q_m is a numerical coefficient of order unity. The assumption is usually justified by the general thermodynamic argument that the quantity $\mathcal{U}_{total} = \mathcal{U}_r + \mathcal{U}_{mag}$ is a minimum as a function of B when $\mathcal{U}_r = (4/3)\mathcal{U}_{mag}$.

In the two cases where reasonably good upper limits on the Compton scattered radiation are available, Centaurus A and the Crab Nebula, $\mathcal{U}_{mag} = \mathcal{U}_r$ to within a factor 2.

If we assume a uniform magnetic field B then Equations (3-63) and (3-65) can be combined to yield

$$B_{eq} = 5\times10^3\ (q_m L/V)^{2/7}\ \nu_*^{-1/7}\ \text{gauss} \qquad (3\text{-}66)$$

$$\mathcal{U}_r = 1\times10^6\ (q_m L/V)^{4/7}\ \nu_*^{-2/7} V\ \text{ergs} \qquad (3\text{-}67)$$

The energies and equipartition magnetic field strengths B_{eq}, as computed by Burbidge (1959) are listed in Table 3.2.

3.9. Changes In The Electron Energy Distribution

In Section 3.7 we saw that a power law electron distribution function produced a power law synchrotron spectrum. The various types of observed spectra from cosmic radio sources are illustrated in Figure 3.3, which shows

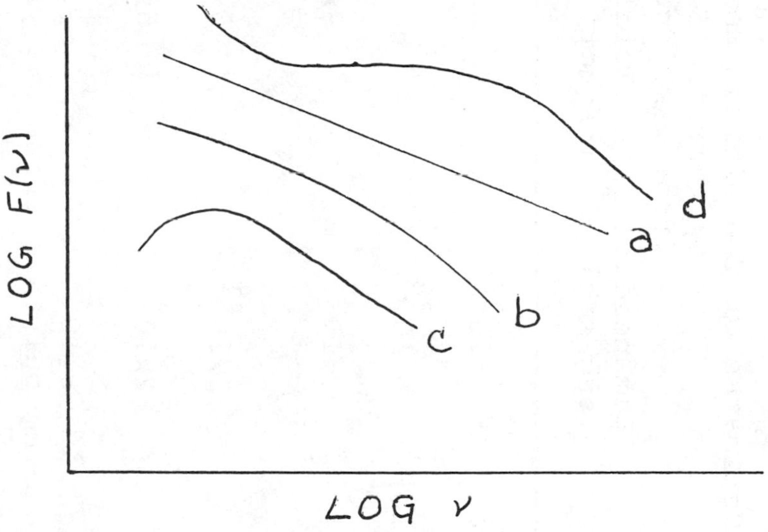

Figure 3.3. Types of observed non-thermal radio spectra.

a graph of log of the flux density, log $F(\nu)$, versus log ν. The spectra of many radio sources are well represented by a single power

Table 3.2

Estimates Of The Minimum Energies Required For Synchrotron Sources (from Burbidge, 1959)

Source	Luminosity (erg/sec)	Total Energy* (magnetic & particles) (ergs)	$\langle B \rangle$* (gauss)
Crab Nebula	10^{36}	$\sim 10^{48}$	$10^{-3} - 10^{-4}$
Galaxy (halo)	$\sim 10^{38}$	$\sim 10^{55}$	$1\text{-}2\times10^{-6}$
M87 (jet)	2×10^{42}	2×10^{54}	2×10^{-4}
Centaurus A (core)	2×10^{41}	3×10^{56}	2×10^{-5}
(halo)	2×10^{41}	5×10^{58}	1×10^{-6}
Cygnus A	6×10^{44}	3×10^{59}	4×10^{-5}

* Assuming only electrons are present

law, over a rather wide range of frequency, or a straight line in Figure 3.3. There are also many sources in which the spectra exhibit curves or bends which can be attributed to changes in the electron energy distribution function. In this section some of the reasons for these changes are discussed. In the next section I discuss synchrotron reabsorption, which is probably responsible for the low frequency turnovers observed in the spectra of many compact sources.

The process by which electrons are accelerated to relativistic energies in a cosmic setting is still not clearly understood. Work on pulsars shows that large scale electromagnetic fields can do it (Ruderman, 1972, and references cited therein), and so can plasma turbulence (Tsytovich, 1973, and references). Here I assume that the acceleration process serves to inject electrons of energy $\gamma m_e c^2$ at a differential rate $Q(\gamma, t)$ into a region where they suffer losses due to synchrotron radiation, Compton scattering and adiabatic expansion. These processes change the energy of relativistic electrons in a well defined way; Synchrotron losses (see Equation (3-6)).

($U_{mag} = B^2/8\pi$, isotropic distribution)

$$(d\gamma/dt)_s = - 3\times10^{-8}\ \gamma^2 U_{mag} \qquad (3\text{-}68)$$

Compton losses (see Equation (4-44) U_{ph} = photon energy density).

$$(d\gamma/dt)_c = - 3\times10^{-8}\ \gamma^2 U_{ph} \qquad (3\text{-}69)$$

Expansion losses

$$(d\gamma/dt)_E = -\gamma v/R = -\gamma/t \tag{3-70}$$

where the second equality holds for uniform expansion (R = source size). The functional form for the expansion losses is obtained by assuming that the non-thermal relativistic electron gas obeys the same adiabatic law as a relativistic Fermi or Bose gas, viz. $\mathcal{U}_r V^{1/3} = \text{const.}$; or by considering the systematic losses suffered by individual relativistic electrons as they collide with expanding magnetic inhomogeneties; or from the conservation of the magnetic moment of the electrons in the expanding magnetic field (see Ginzburg and Syrovatskii, 1964). In all the processes (3-68), (3-69), (3-70) the energy of the electron changes gradually and the change in the distribution function $N_r(\gamma, t)$ can be described by a continuity equation in energy space:

$$\partial N_r/\partial t + (\partial/\partial\gamma)(\mathring{\gamma} N_r) = Q(\gamma, t) \tag{3-71}$$

where Q represents sources and sinks of relativistic electrons corresponding to injection into the source from an accelerating region or gradual leakage from the source. Often the leakage loss is expressed explicitly by a term $- N_r\ t_d$ where t_d is the characteristic time for diffusion out of the source. If in addition the volume of the source is changing in divergence term $\text{div}N\vec{v}$ must be added to the left hand side. An equivalent and often simpler formulation of the continuity equation is the integral form:

$$N_r(>\gamma, t)V(t) = \int_\gamma^\infty V(t)N_r(\gamma', t)$$

$$= V(t) \int_{\gamma(t_o)}^{\infty} N_r(\gamma, t_o) d\gamma + \int_{t_1}^{t} dt' V(t')$$

$$\cdot \int_{\gamma(t')}^{\infty} Q(\gamma', t') d\gamma' \qquad (3\text{-}72)$$

where $N_r(> \gamma, t)$ is the number of particles with energies greater than $\gamma m_e c^2$, $V(t)$ is the volume of the source at time t, and t_o refers to some initial instant.

For stationary distribution functions $\partial N/\partial t = 0$ and the solution of equation (3-71) has the form

$$N(\gamma) = (d\gamma/dt)^{-1} \int_{\gamma}^{\infty} Q(\gamma') dt' \qquad (3\text{-}73)$$

For a power law injection spectrum of the form $Q(\gamma) = Q_i \gamma^{-n}$,

$$N(\gamma) = Q_i \gamma^{1-n}/(d\gamma/dt)(n-1) \qquad (3\text{-}74)$$

For synchrotron losses $(d\gamma/dt) \propto \gamma^2$, so

$$N(\gamma) \propto \gamma^{-(1+n)} \qquad (3\text{-}74')$$

This will occur for particles whose energy is such that their radiative lifetime is less than the age of the source t, $(t_{sy} \ll t)$. For $t_{sy} > t$ radiation losses have no effect and the spectrum retains its initial form.

In terms of the spectrum of the emitted radiation, a "break" or bend in the spectrum will occur at a frequency given approximately by (see Equation (3-9)).

$$\nu_b = 10^{24}/B_\perp^3\ t^2 \qquad (3\text{-}75)$$

For $\nu << \nu_b$, the spectral index is given by (3-49). For $\nu >> \nu_b$, $q = n/2$ i.e., the spectrum steepens by one half power.

The intensity of the radiation remains constant for frequencies below the break, since there is a balance there between synchrotron losses and particle injection.

Below the break frequency which varies with time according to (3-75), no such balance exists and the intensity increases linearly with time for injection at a constant rate.

Another case of interest is obtained by assuming that the injection occurs in a very short time and ceases thereafter. We can obtain a solution in this case using (3-72). In the absence of sources or sinks, Equation (3-72) becomes

$$N(\gamma,t)d\gamma = N(\gamma_o)d\gamma_o = N_o\gamma_o^{-n}\ d\gamma_o \qquad (3\text{-}76)$$

where the subscripts denote the conditions just after the injection. For a particle suffering synchrotron losses we have

$$\gamma = \gamma_o/(1 + b\gamma_o t) \qquad (3\text{-}77)$$

where $b = 3\times10^{-8}\ U_{mag}$. From (3-76) and (3-77) we obtain, expressing γ_o in terms of γ and t,

$$N(\gamma,t) = N(\gamma_o)\frac{d\gamma_o}{d\gamma} = \frac{N_o\ \gamma^{-n}}{(1-b\gamma t)^{2-n}} \qquad \gamma_1' < \gamma < \gamma_2'$$

$$= 0 \qquad \gamma < \gamma_1' \text{ or } \gamma > \gamma_2' \qquad (3\text{-}78)$$

where $\gamma' = \gamma/(1 + b\gamma t)$ and

$$N(\gamma,0)d\gamma = \begin{cases} N_o\gamma^{-n}d\gamma & \gamma_1 < \gamma < \gamma_2 \\ 0 & \gamma < \gamma_1 \text{ and } \gamma > \gamma_2 \end{cases} \quad (3\text{-}79)$$

Thus even with an initial energy distribution extending to unlimited energy, there will be a cutoff at $\gamma = 1/bt$.

If the electron trajectories are initially distributed over a range of angles, the particles with small pitch angles lose energy slowly (see Equation (3-5), and (3-8)). There is no sharp cutoff energy but rather a break in the energy spectrum, and consequently the radiation spectrum. For $\nu \ll \nu_b$ given by (3-75) the radiation losses will not be important and $q = (n-1)/2$. For an initially isotropic pitch angle distribution and γ_2 sufficiently high, the spectrum above the break frequency will be a power law with an index $q^1 = (4/3)q + 1$.

However, it is likely that in cosmic sources, the particles with small pitch angles will be scattered to large pitch angles by plasma instability or irregularities in the magnetic field, so that the spectrum above the break will be steeper than $(4/3)q + 1$.

Compton losses produce exactly the same effect as synchrotron losses, with the magnetic energy density U_{mag} replaced by U_{ph} in (3.77). It is probable that curved spectra of type b in Figure 3.4 are produced by either synchrotron or Compton losses.

An intermediate case between the two

extremes considered above is one in which particles are injected through a series of recurrent bursts. The subsequent synchrotron radiation spectrum will then vary with time in the following manner: at a time t after the first injection, the effect of radiation losses will not be important for $\nu << \nu_b$ and the spectral index will be equal to its initial value q. In the frequency range $\nu_b \lesssim \nu \lesssim \nu_b(t/\tau)^2$, the time scale for electron energy losses is longer than the period τ between the outbursts and the injection can be considered to be quasi-continuous. The spectrum is then in equilibrium and the spectral index is q + 1/2. For $\nu \gtrsim \nu_b(t/\tau)^2$ the rate of injection is not sufficient to balance the radiation losses and the spectrum steepens to a (4/3)q + 1 or greater. However, if the source is observed soon after an outburst, the high frequency spectrum will be dominated by the young electrons which have not yet decayed and the spectrum will flatten to q up to a frequency $\nu \sim \nu_b(t/t')^2$ where it then becomes $(4/3)q_o + 1$ or steeper (t′ is the time since the last outburst).

Kellerman (1966) has interpreted the observed radio source spectra in terms of a recurrent injection model. The flattest spectra observed have indices near 0.25. This suggests that the spectral index at the time of injection is approximately 0.25 corresponding to an index for the energy distribution n_o = 1.5. If the injection were instantaneous, the final index would be (4/3)q + 1 = 1.33 which is close to the steepest spectra observed. If the electrons are continuously replenished then q = 0.75, near the mean index of 0.77 for the several hundred non-thermal radio sources considered by Kellerman.

Thus, according to this model, sources having spectra with $q \simeq 0.25$ may be interpreted as young sources where synchrotron losses have not yet taken their toll. The steepest spectra then belong to sources where either only a single burst has occurred or where the period between bursts is sufficiently great that the spectra have had time to reach the limiting value of $(4/3)q + 1 \simeq 1.33$. In the majority of sources where $q \simeq 0.75$ over a wide range of frequencies, the relativistic particles are produced sufficiently often that radiation losses are in equilibrium with particle injection.

Another process that can effect a change in the intensity of the radiation is the expansion of the source. If no additional energy is pumped into the relativistic particles and there is no injection of new particles into the sources, the energy of the relativistic particles changes according to Equation (3-74) so $\gamma \propto R^{-1}$ where R is the source radius. The total number of relativistic particles in the source does not change so

$$V \int_{\gamma_1}^{\gamma_2} N(\gamma)d\gamma = \text{const.} \qquad (3\text{-}80)$$

Assuming $N(\gamma) = N_r\gamma^{-n}$ with $n > 1$ and $\gamma_2 >> \gamma_1$, and using the fact that $\gamma \propto 1/R$, it follows from (3-80) that $N_rV \propto R^{1-n}$. From Equations (3-50) and (3-60), the spectral luminosity of the synchrotron source varies as

$$L(\nu) \propto B^{(n+1)/2} R^{1-n} \propto R^{1-n-m(n+1)/2} \quad (3\text{-}81)$$

for $B \propto R^{-m}$. For expansion with flux conservation, m = 2; for pulsar generated fields m = 1 (see Ruderman 1972 and references). For a nebula expanding at a rate v = dR/dt, the fractional change in the spectral luminosity will be

$$(dL(\nu)/dt)/L(\nu) = (2-2n-mn-m)v/2R \quad (3\text{-}82)$$

This effect was predicted by Shklovsky in 1960 and observed for the radio source Cas A which is an expanding supernova remnant. The observations of Cas A (Kellerman and Pauliny-Toth and references, 1968) imply that

$(dL/dt)/L \sim -3\times10^{-10}$ and $v/R \sim 10^{-10}$, and $n = 2.5$, so if the model (3-82) applies, then $m \sim 1$, corresponding to a pulsar generated magnetic field. However, other effects such as continued injection of particles and departure from spherical symmetry complicate the radio evolution (Woltjer, 1972) so the conclusions based on this simple model cannot be taken as definite.

3.10. Synchrotron Self-Absorption

The reabsorption of synchrotron radiation is important for compact sources. To estimate its importance, we note that for sources optically thick to synchrotron reabsorption, the energy of the electrons γmc^2 should be related to the brightness temperature of the source by

$$KT_B \sim \gamma mc^2 \quad (3\text{-}83)$$

The brightness temperature T_B is the temperature a source would have if it emitted the same spectral flux at a given frequency according to the Rayleigh-Jeans law; it is therefore related to the spectral flux density $F(\nu)$ ergs/(cm^2 sec Hz) by

$$2KT_B \sim (c/\nu)^2 \; F(\nu)(d/R)^2/\pi \qquad (3\text{-}84)$$

where d is the distance to the source and R is the radius of the source. Then using Equation (3-7) for the typical synchrotron frequency, one can eliminate γ and T_B to obtain a criterion for synchrotron self-absorption:

$$R \lesssim 10^{15} \; F(\nu)^{1/2} \; B^{1/4} \; \nu^{-5/4} \; d \quad \text{cm.} \qquad (3\text{-}85)$$

In terms of the angular diameter θ subtended by the source

$$\theta \lesssim 4\times10^{-3} F_{-26}{}^{1/2} \; B^{1/4} \; \nu_8{}^{-5/4} \; \text{arc sec} \qquad (3\text{-}86)$$

where $F_{-26} = F\times10^{26}$ erg/cm^2 sec Hz, and $\nu_8 = \nu/10^8$ Hz.

Synchrotron self absorption may be the cause of the low frequency turnovers observed in spectra (see Fig. 3.4). Other possibilities are the cyclotron turnover (3-52) which is due to small pitch angle effects, the Razin-Tsytovich effect (see next section) and free-free absorption (see Chapter 5).

To compute the coefficient for reabsorption of synchrotron radiation by ultrarelativistic electrons we make use of the fact that, in the ultrarelativistic limit

$$W = pc \tag{3-87}$$

If we transform from an isotropic electron momentum distribution to an electron energy distribution according to

$$4\pi N(p)\, p^2\, dp = N(\gamma)d\gamma \tag{3-88}$$

and substitute into (2-158) we find

$$\mu(\nu) = (-1/8\pi\ m_e\nu^2) \int P(\nu,\gamma)\gamma^2 (\partial/\partial\gamma) \cdot (N(\gamma)/\gamma^2)d\gamma \tag{3-89}$$

where we have set $\int(dP/d\Omega)d\Omega = P(\nu,\gamma)$.

The expressions for the synchrotron absorption coefficient for a power law distribution of electrons is given in Ginzburg and Syrovatskii (1969) and Pacholczyk (1970). In order of magnitude we have

$$\mu(\nu) \sim (1/8\pi\ m_e\nu^2)P\ N(\gamma)/\nu \sim (1/8\pi\ m_e\nu^2)j(\nu)(\nu_B/\nu)^{\frac{1}{2}} \tag{3-90}$$

(j = emissivity = power emitted per unit volume). The optical depth for synchrotron absorption is

$$\tau(\nu) \sim \mu(\nu)N(\gamma)\gamma R \sim (1/8\pi\ m_e\nu^2) \cdot (\nu_B/\nu)^{\frac{1}{2}}\ I(\nu)\ N_r \tag{3-91}$$

where $I(\nu)$ is the spectral intensity of the radiation at the source.

For a power-law electron distribution function with index *n*, the polarization of an optically thick source is (Ginzburg and Syrovatskii, 1969)

$$\Pi = 3/(6n + 13) \tag{3-92}$$

3.11. The Razin Effect

This is another effect which can cause a low frequency cut off in synchrotron spectra. It has its basis in the fact that in a plasma the index of refraction is less than unity, so that the phase velocity of light in the plasma is greater than c. Consequently, an electron may be moving at a speed almost equal to c and still be moving at a velocity much less than the phase velocity of low frequency waves. For these waves the relativistic beaming and amplification of the radiation will not occur. Formally this effect can be understood by replacing the factor $1 - \beta \cos\theta$ which appears in the relativistic radiation formulas by $(1-n_r\, \beta \cos\theta)$ with $n_r = (1 - \nu_p^2/\nu^2)^{\frac{1}{2}} \sim 1 - \nu_p^2/2\nu^2)$ (ν_p is the plasma frequency; see Sections 1.10 and 1.11). Then, using the expansion $\beta \simeq 1 - 1/2\, \gamma^2$, $\cos\theta \simeq 1 + \theta^2/2$, we find

$$(1 - n_r\, \beta \cos\theta) \sim \nu_p^2/2\nu^2 + 1/2\, \gamma^2 - \theta^2/2 \tag{3-93}$$

The expression (3-93) will be small for

$$\theta \sim 1/\gamma(1 + \nu_p^2\gamma^2/\nu^2)^{\frac{1}{2}} \tag{3-94}$$

Therefore for frequencies such that $\nu_p^2\, \gamma^2 \gtrsim \nu^2$ the plasma will affect the beaming of the radiation. Since most of the radiation is emitted near the critical frequency $\nu_c = 3\nu_B\,\gamma^2/2$, the effect of the plasma on synchrotron emission will be important for frequencies such that

$$\nu \lesssim \nu_p^2/\nu_B = 20\ N_e/B \qquad (3\text{-}95)$$

This effect was first applied to astrophysical problems by V. Razin.

The expression for the emitted spectrum is obtained by making the substitution (3-93) in the radiation formulas (see Section 1.11) with the result that (cf. Equation (3.37))

$$P(\nu) = (2\pi\sqrt{3}^2\ \nu_B \sin\alpha/c)(1 + \nu_p^2\gamma^2/\nu^2)^{-\frac{1}{2}} \cdot F(\nu/\nu_c^1) \qquad (3\text{-}96)$$

where

$$\nu_c^1 = \nu_c(1 + \nu_p^2\ \gamma^2/\nu^2)^{-3/2} \qquad (3\text{-}97)$$

References

Detailed discussions of all aspects of synchrotron radiation can be found in

Jackson (1962), Chapter 14

Ginzburg and Syrovatskii (1965) and (1966)

Blumenthal and Gould (1970)

Bekefi (1966), Chapter 6

Pacholczyk (1970), Chapter 3

Landau and Lifshitz (1962), Chapter 9

In addition, the following references contain discussions of specific topics:

Small Pitch Angle Effects:

Epstein (1973)

Kinetic Equation And Changes In Electron Distribution:

Ginzburg and Syrovatskii (1964)

Melrose (1969)

Kardshev (1962)

Scheur and Williams (1968)

Problems

3.1. As cosmic ray electrons and protons traverse the interstellar magnetic field they lose energy by synchrotron radiation. Assuming that the average strength of the component of the magnetic field perpendicular to the direction of the charge's motion is 10^{-6} gauss, compute the limiting energies which cosmic ray electrons and protons can have after traversing a distance of 30,000 light years through this field.

3.2. The Crab Nebula is observed to emit X-rays having an energy of at least 100 keV from an extended region. Compute the energy, lifetime and Larmor radius of the electrons producing this radiation, assuming that it is due to synchrotron radiation in a magnetic field of 10^{-4} gauss.

3.3. The nonthermal spectrum of the Crab Nebula exhibits a bend around 10^{15} Hz. Assuming that this bend is due to synchrotron losses, use the known age of the Crab of $\simeq$ 900 years to estimate the magnetic field strength.

3.4. The observations of the time variations of the pulsar in the Crab Nebula imply that the optical pulsar beam (frequency = 6×10^{14}Hz) has an angular structure of less than 6×10^{-3} radians. Assuming that the radiation pattern is due to a collection of relativistic electrons with a characteristic energy $\overline{W}$ streaming along a magnetic, what restrictions are placed on the characteristic energy and pitch angle of the electrons? On the magnetic field strength in the radiating region? Can small pitch angle effects account for the observed turnover in the spectrum around 6×10^{14} Hz? (Epstein and Petrosian, 1973, Ap. J. 183, 611)

3.5. Estimate the energy in relativistic electrons and the equipartition magnetic field for the following sources.

(1) Crab Nebula

$L(\nu) = 4\times10^{22}$ erg/cm^2 sec Hz at $\nu = 10^{15}$ Hz

$L(\nu) \propto \nu^{-1}$ between 10^{15} and 10^{20} Hz

$L(\nu) \propto \nu^{-0.25}$ between 10^{8} and 10^{15} Hz

volume = 3×10^{56} cm^3

(2) Cygnus A

$L(\nu) = 3\times10^{35}$ erg/cm^2 sec Hz at $\nu = 10^{9}$ Hz

$L(\nu) \propto \nu^{-0.8}$ between 10^{8} and 10^{9} Hz

$L(\nu) \propto \nu^{-1.2}$ between 10^{9} and 10^{10} Hz

volume = 3×10^{68} cm^3

(3) Typical QSO

$L(\nu) = 10^{31}$ erg/cm^2 sec Hz, at $\nu = 10^{15}$ Hz

flat spectrum between 10^{14} and 10^{15} Hz

volume = 3×10^{51} cm^3

3.6. Consider a source that is optically thick at a frequency below 10^{11} Hz at time t. If the source expands at a constant rate and the magnetic field changes as R^{-2} with the expansion, show that the flux density in the optically thick region changes as

$$S(\nu, t) = S(\nu, t_1)(t/t_1)^3$$

Use Equations (3-82) and (3-86) to describe evolution of the entire spectrum (movement of maximum, temporal behavior in optically thin region) as the source expands. Assume that the spectral index r in the optically thin region is 0.75. (See Kellerman and Pauliny-Toth, 1968, Annual Rev. Astr. Ap. 6, 417.)

Chapter 4

ELECTRON SCATTERING

4.1. Scattering By Free Charges At Rest--The Thompson Formula

If an electromagnetic wave falls on a system of charges, the charges will be accelerated and emit radiation in all directions. For nonrelativistic motion, the frequency of this radiation will be the same as that of the incident radiation. The incident wave is said to have been scattered.

The scattering is most conveniently characterized by the differential cross section, which is defined as the time average of the ratio of the amount of energy emitted by the scattering system in a given direction per unit time to the energy flux density of the incident radiation.

Differential Cross Section =

$$\frac{d\sigma}{d\Omega} = \frac{\text{Energy radiated/unit time/unit solid angle}}{\text{Incident energy flux density}}$$

$$= \left(\frac{dP/d\Omega}{S}\right)_t \tag{4-1}$$

The incident energy flux S is just the time averaged Poynting's vector for the incident wave:

$$S = c\,|E_o|^2/8\pi \tag{4-2}$$

Here we have assumed that the incident wave is a plane monochromatic linearly polarized wave

$$\vec{E} = \vec{E}_o\, e^{i(\vec{k}\cdot\vec{r} - \omega t)} \tag{4-3}$$

Consider the scattering produced by a free charge at rest.

Assume that the velocity acquired by the charge under the influence of the electric field of the incident wave is small compared with the velocity of light. Then the magnetic forces acting on the charge can be neglected in comparison to the electric force eE. We can also neglect the effect of the displacement of the charge during the vibrations under the influence of the field, and we can assume that the field which acts on the charge at all times is the same as that at the origin:

$$\vec{E} = \vec{E}_o\, e^{-i\omega t} \tag{4-4}$$

The equation for the dipole moment of the charge is then

$$\ddot{\vec{d}} = (e^2/m_e)\vec{E} \tag{4-5}$$

Inserting this into the formula (1-139) we obtain

$$dP/d\Omega = r_o^2 c \sin^2\Theta\ |E|^2/4\pi \tag{4-6}$$

where $r_o = e^2/m_e c^2$ is the <u>classical electron radius</u>.

Together with (4-1) and (4-2) this yields the differential cross section for the scattering by a free charge at rest:

$$d\sigma/d\Omega = r_o^2 \sin^2\Theta \tag{4-7}$$

independent of frequency. Remember Θ is the angle between the direction of observation $\hat{n}$ and $\ddot{\vec{d}}$ or $\vec{E}$. Integrating over solid angle, we find the total cross section:

$$\sigma_T = 8\pi\, r_o^2/3 \qquad (4\text{-}8)$$

This is called the Thomson cross section.

For unpolarized radiation, we must average over all directions of E in a plane perpendicular to the direction of propagation of the incident wave. If n is the direction of the scattered radiation, then (see Figure 4-1).

$$\hat{n} \circ \vec{E} = E \sin\theta \cos\varphi = E \cos\Theta \qquad (4\text{-}9)$$

and

$$(\hat{n} \cdot \vec{E})^2 = E^2 \sin^2\theta \cos^2\varphi \qquad (4\text{-}10)$$

Averaging over φ.

$$\langle (\hat{n}\cdot\vec{E})^2 \rangle_\varphi = (1/2)E^2 \sin^2\theta = E^2\, \overline{\cos^2\Theta} \qquad (4\text{-}11)$$

Therefore

$$\overline{\sin^2\Theta} = (1 - \overline{\cos^2\Theta}) = (1 + \cos^2\theta)/2 \qquad (4\text{-}12)$$

and for unpolarized incident radiation

$$d\sigma/d\Omega = r_o^2\,(1 + \cos^2\theta)/2 \qquad (4\text{-}13)$$

where θ is the scattering angle, i.e., angle between the incident and scattered wave.

The classical result (4-13) is valid only when the frequency $\nu << m_e c^2/h$. When the

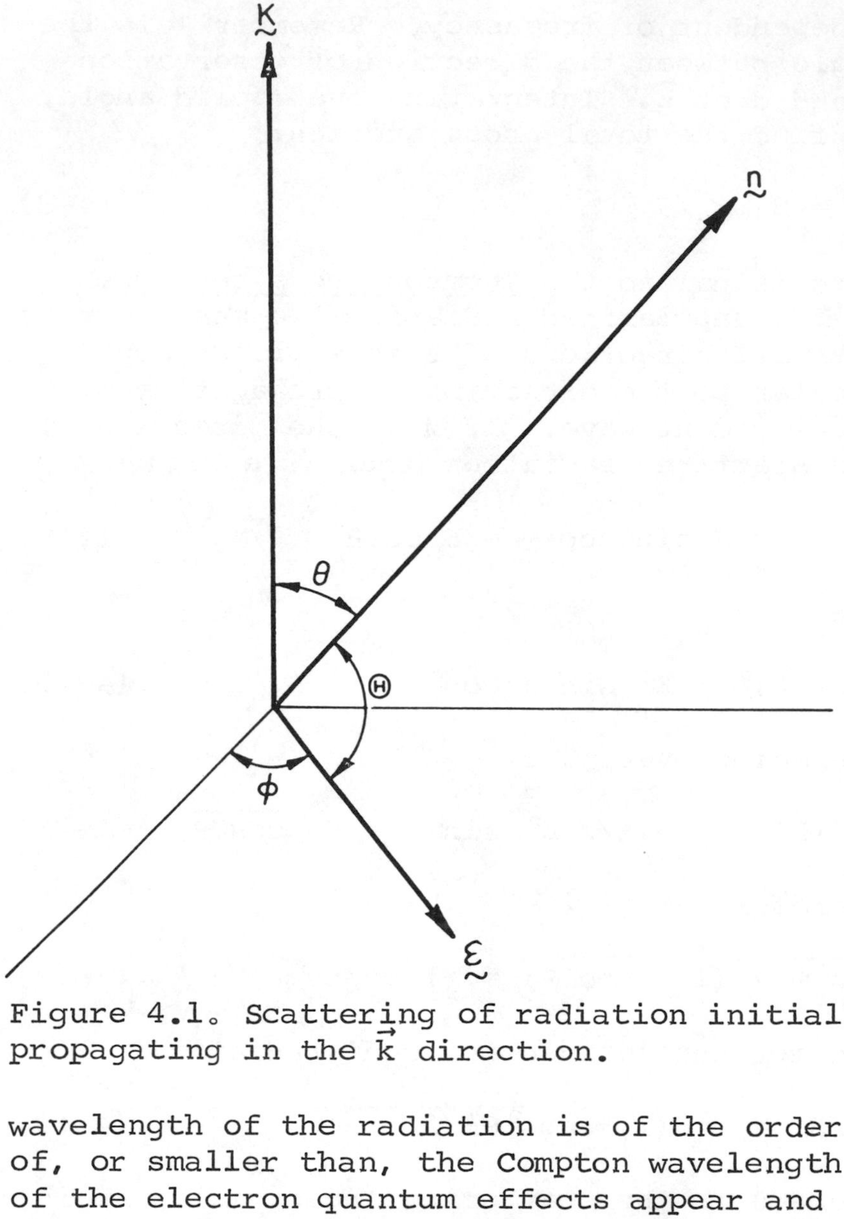

Figure 4.1. Scattering of radiation initially propagating in the $\vec{k}$ direction.

wavelength of the radiation is of the order of, or smaller than, the Compton wavelength of the electron quantum effects appear and the frequency of the scattered wave is <u>not</u>, in general, equal to the frequency of the incident wave. The general process of scattering of waves of arbitrary frequency by a free charge is called Compton scattering. It is described by the Klein-Nishina formula which is the same as the Thomson formula at low frequencies and decreases rapidly at high frequencies $h\nu \gg m_e c^2$. For protons the departures from the Thomson formula occur at photon energies of the order of the rest energy of the pi meson.

4.2. Scattering By Free Charges At Rest--The Compton And Klein-Nishina Formulas

Consider the following process: a primary light quantum with momentum $\hbar\vec{k}_i$ collides with a free electron assumed to be initially at rest. Its energy and momentum are given by

$$\vec{p}_i = 0 \qquad W_o = m_e c^2 \tag{4-14}$$

The general case $\vec{p}_i \neq 0$ can be obtained by a Lorentz transformation. In the final state the photon has momentum $\hbar\vec{k}_f$, and energy $h\nu$. Since total momentum is conserved, the final state of the electron has a momentum

$$\vec{p} = \hbar(\vec{k}_i - \vec{k}_f) \tag{4-15}$$

Conservation of energy requires

$$W + h\nu = m_e c^2 + h\nu_i \tag{4-16}$$

Denoting the angle between the incident and scattered wave by θ we obtain

$$\nu = \frac{\nu_i}{1 + (h\nu_i/m_e c^2)(1- \cos\theta)} \tag{4-17}$$

which is the <u>Compton formula</u> for the shift in

frequency of the scattered radiation. In the non-relativistic case $h\nu_i << m_e c^2$ and $\nu = \nu_i$.

In the extreme relativistic case $h\nu_i >> m_e c^2$ and the frequency of the scattered photon depends on the scattering angle. For very small angles such that

$$h\nu_i \, (1 - \cos\theta) << m_e c^2$$

or

$$\theta << (2m_e c^2 / h\nu_i)^{\frac{1}{2}} \qquad (4\text{-}18)$$

we have $\nu \simeq \nu_i$. For large angles, i.e., $\theta >> (2\, m_e c^2 / h\nu_i)^{\frac{1}{2}}$,

$$\nu = m_e c^2 / h(1 - \cos\theta) \qquad (4\text{-}19)$$

and the energy of the scattered quantum is always on the order of $m_e c^2$ independent of the primary frequency. The wavelength is of the order of the Compton wavelength λ_c

$$\lambda = c/\nu = h/m_e c \equiv \lambda_c \qquad (4\text{-}20)$$

To compute the transition probability from the initial state $(\vec{k}_i, \vec{p}_i)$ to the final state $(\vec{k}, \vec{p})$, we have to use second order perturbation theory since the process can happen only by passing through an intermediate state which can differ by one quantum only from the initial and final states. Since for these intermediate states momentum (but not energy) is conserved, the following two intermediate states are the only ones possible: (i) the incoming photon is absorbed by the electron and then the electron emits the outgoing photon, and (ii) the electron emits a photon and subsequently absorbs the incident photon.

An electron moving with relativistic velocity with a given momentum $\vec{p}$ can exist in four states, corresponding to the fact that the electron may have either of two spin directions and also a positive or negative energy

$$W = \pm (p^2 c^2 + m_e^2 c^4)^{\frac{1}{2}} \quad (4\text{-}21)$$

All these four states must be taken into account as intermediate states. Each of the two intermediate states mentioned above is therefore actually fourfold because only the momentum is determined. On the other hand, in the initial and final states the electron has a positive energy and a given spin direction.

An alternative way of looking at the intermediate states is in terms of the creation of virtual pairs of positrons and electrons. Thus in process (ii) $\hbar\vec{k}_i$ creates a pair $\vec{p}^+ = \hbar\vec{k} - \vec{p}_i$, $\vec{p}^- = \vec{p}$. In the transition to the final state $\vec{p}_i$ and $\vec{p}^+$ annihilate with the emission of $h\vec{k}$.

Upon evaluating the matrix element summed over final spin states and averaged over initial spin states, the differential cross section is found to be (see Heitler, 1954)

$$\frac{d\sigma}{d\Omega} = \frac{1}{4} r_o^2 \left(\frac{\nu}{\nu_i}\right)^2 \left[\frac{\nu_i}{\nu} + \frac{\nu}{\nu_i} - 2 + 4 \cos^2 \Theta\right] \quad (4\text{-}22)$$

which is the Klein-Nishina formula. Here Θ

is the angle between the directions of polarization of the incident and scattered photons. Averaging over initial polarizations and summing over final polarizations,

$$\frac{d\sigma}{d\Omega} = \frac{1}{2} r_o^2 \left(\frac{\nu}{\nu_i}\right)^2 \left(\frac{\nu_i}{\nu} + \frac{\nu}{\nu_i} - \sin^2\theta\right) \qquad (4\text{-}23)$$

The total scattering cross section is

$$\sigma = 2\pi r_o^2 \left\{ \frac{1+\Gamma}{\Gamma^3} \left[\frac{2\Gamma(1+\Gamma)}{1+2\Gamma} - \log(1+2\Gamma)\right] \right.$$

$$\left. + \frac{1}{2\Gamma} \log\,(1+2\Gamma) - \frac{1+3\Gamma}{(1+2\Gamma)^2} \right\} \qquad (4\text{-}24)$$

where

$$\Gamma = h\nu_i/m_e c^2$$

In the non-relativistic case

$$\sigma = \sigma_T(1 - 2\Gamma + (26/5)\Gamma^2 + \ldots \qquad (4\text{-}26)$$

where σ_T is the classical cross section for Thomson scattering (see Equation (4-8)).

In the extreme relativistic case,

$$\sigma = (3\sigma_T/8)\Gamma^{-1}\ (\log 2\Gamma + 1/2) \qquad (4\text{-}27)$$

Thus Compton scattering is negligible at high frequencies where pair production become important (see Section 4.7).

4.3. Radiation Pressure

In the scattering process the electromagnetic wave exerts a force on the scattering electron. Because the scattering cross section (4-13) is symmetric between forward and backward angles, on the average the wave incident on the electron loses momentum at a rate of $(\overline{U}/c)\sigma c$ where $\overline{U}$ is the average energy density in the wave. Since the total flux of momentum of the scattered wave is zero in the nonrelativistic case, this implies that the lost momentum is "absorbed" by the scattering electron. The average force acting on the electron is therefore

$$\overline{F}_{rad} = \sigma_T \overline{U} \qquad (4\text{-}28)$$

in the direction of propagation of the incident wave.

The pressure exerted by a radiation field on a gas is just the force per unit area:

$$P_{rad} = \sigma_T \overline{U}\, \lambda_{mfp} N_e = \overline{U} \qquad (4\text{-}29)$$

where λ_{mfp} is the mean free path for the photon to scatter off an electron $= (N_e \sigma_T)^{-1}$ and it is assumed that the radiation is moving normal to the surface. For the more general case

$$P_{rad} = (1/4\pi) \int \overline{U}(\theta) \cos^2\theta \, d\Omega \qquad (4\text{-}29')$$

which for an isotropic radiation field reduces to

$$P_{rad} = (1/3)\, \overline{U} \qquad (4\text{-}29'')$$

For an equilibrium radiation field $\overline{U} \propto T^4$ so $P_{rad} \propto T^4$. Therefore an equilibrium radiation field behaves like a gas with a ratio of specific heats $c_p/c_v = 5/3$.

This fact is important for astrophysical applications, because it limits the amount of power that can be generated by normal stars or by accretion of matter under the influence of gravitational forces.

In accretion, gravitational energy is released by particles falling onto the surface of a star or other massive body. A particle of mass m gains an amount of kinetic energy $\sim GMm/R$ in falling from infinity onto the surface of a star of mass M, and radius R. Through interaction with the atmosphere of the star this energy is transformed into radiant energy directly by means of bremsstrahlung, and indirectly by means of heating of the atmosphere followed by thermal radiation. Accretion has been postulated as the energy source for such phenomena in astrophysics, as cosmic X-ray sources and quasi-stellar sources.

Equation (4-28) shows that the process will stop when the energy output reaches the point where the energy density in the electromagnetic field is so high that the radiation forces offset the gravitational forces. To evaluate this limiting luminosity correctly we must take account of the fact that both the radiation and the gravitational forces are different for the electrons and the protons. Therefore, the charges will tend to separate under the action of these forces and an electric field will be set up to oppose this separation. Very quickly the system will reach an equilibrium situation in which the total force acting on the electrons and protons is the same, so that they move together. This enables us to obtain the induced electric field E_i.

The forces acting on the electrons and protons are

$$F_e = (GMm_e/r^2) - \sigma_T\overline{U} - eE_i \qquad (4\text{-}30)$$

and

$$F_p = (GMm_p/r^2) - \sigma_T(m_e/M_p)^2 + eE_i \qquad (4\text{-}30')$$

respectively. The induced electric field is obtained by setting $F_e = F_p$:

$$eE_i = -\frac{1}{2}((GMm_p/r^2) + \sigma_T\overline{U}) \qquad (4\text{-}31)$$

where we have neglected terms of order (m_e/m_p) and higher. The net force on an electron or proton is therefore

$$F_e = F_p = \frac{1}{2}((GMm_p/r^2) - \sigma_T\overline{U}) \qquad (4\text{-}32)$$

The limiting radiant energy density beyond which the accretion process stops and turns into expansion under the action of radiation pressure is

$$\overline{U} = GMm_p/r^2\sigma_T \qquad (4\text{-}33)$$

Defining the energy density $\overline{U}$ in terms of the luminosity of the source L through the relation (see Equation (2-29))

$$\overline{U} = L/4\pi r^2 c \qquad (4\text{-}34)$$

we find that the maximum luminosity that can be attained in an accretion process is

$$L_{max} = 4\pi GMm_p c/\sigma_T = 1\times 10^{38}\ (M/M_\odot)\ \text{erg/sec} \tag{4-35}$$

This is the Eddington limit.
It also determines the maximum luminosity for stable stars.

Radiation pressure is not caused exclusively by scattering of photons by electrons. In many situations the absorption of momentum by certain stoms or ions is a greater source of pressure than electrons and can drive stellar winds and perhaps even galactic winds (see Weymann, 1974, for a review).

4.4. Scattering of Radiation By Moving Electrons

The scattering of an electromagnetic wave by moving electrons differs from the scattering by charges at rest in that the frequency of the scattered radiation can be greater than the frequency of the incident wave. In the system of coordinates in which the charge is at rest, a wave of frequency ν_i appears as a Doppler shifted wave of frequency

$$\nu_i' = \gamma\nu_i\ (1 - \beta\cos\psi) \tag{4-36}$$

Here $\gamma = (1-\beta^2)^{-\frac{1}{2}}$, $\beta = v/c$, v is the velocity of the electrons, and ψ is the angle of incidence in the laboratory system.

In the reference frame moving with the charge, we can use the Thomson formula (4-13) to describe the scattering for photons satisfying the inequality $h\nu_i << m_e c^2/\gamma$. According to this formula, the frequency of the wave is unchanged in the scattering, so

$$\nu'_f = \nu'_i \qquad (4\text{-}37)$$

Transforming back to the laboratory system, the frequency of the scattered wave is found to be

$$\nu_f = \gamma^2 \nu_i (1 - \beta\cos\psi)(1 + \beta\cos(\theta' + \psi')) \qquad (4\text{-}38)$$

where ψ' and θ' are the angle of incidence and the scattering angle, respectively, in the moving reference frame.

Equation (4-38) shows that, for ultrarelativistic electrons ($\gamma >> 1$), we will in general have

$$\nu_f \sim \gamma^2 \nu_i \qquad (4\text{-}39)$$

Thus very high frequencies can be produced by scattering from relativistic electrons. This situation, where the wave gains energy from the electron, is the inverse of the usual Compton scattering. Hence, it is often called the inverse Compton effect. This mechanism has been used in connection with theories to explain the cosmic x-ray background in terms of scattering the microwave background radiation by cosmic ray electrons (see Felten and Morrison, 1966, Silk, 1973).

The scattered power is invariant, so

$$P = \int 2\pi\, (\nu_f/\nu'_i)\sigma(\theta')\sin\theta'\ d\theta'\ I(\nu'_i, \psi)d\nu'_i \qquad (4\text{-}40)$$

where $I(\nu'_i, \psi')$ is the intensity of the incoming beam, and $\sigma(\theta')$ is given by (4-13).

Using (4-36) and (4-38), and the fact that time dilation cancels a factor γ, (4-40) becomes

$$P(\psi) = \pi r_0^2 \, I(\psi)\gamma^2 (1-\beta\cos\psi)^2 \int_o^{\pi} (1+\cos^2\theta')$$

$$\cdot \; (1 + \beta\cos(\theta' + \psi')\sin\theta' \, d\theta' \qquad (4\text{-}41)$$

where

$$I(\psi) = \int I(\nu,\psi)d\nu \qquad (4\text{-}42)$$

For ultrarelativistic electrons ($\gamma >> 1$) $\psi' \sim \pi$, independent of ψ. Equation (4-41) then becomes

$$P(\psi) = (8\pi/3) \; r_0^2 \; I(\psi)\gamma^2 (1-\beta\cos\psi)^2$$

$$\cong 32\pi/3 \; r_0^2 \; I(\psi)\gamma^2 \; \sin^4\psi/2 \qquad (4\text{-}43)$$

For an isotropic radiation field, $I = cU_{ph}/4\pi$, so

$$P = (32\pi/9) \; r_0^2 c \; U_{ph} \; \gamma^2$$

$$= 2.6\times10^{-14} \; U_{ph} \; \gamma^2 \quad \text{erg/sec} \qquad (4\text{-}44)$$

This expression is identical with (3-6) for synchrotron losses with the photon energy density replacing the energy density of the isotropic magnetic field. Thus, the Compton losses will be more important than synchrotron losses when $U_{ph} > B^2/8\pi$.

The time for a relativistic electron to lose a significant fraction of its energy by

Compton scattering is

$$t_c \sim m_e c^2 \gamma / P \sim 3\times10^7/U_{ph}\gamma \quad \text{sec} \tag{4-45}$$

For the microwave background $U_{ph} \sim 10^{-13}$ erg/cm^3, so Compton losses have little effect on the energy spectrum of the electrons except at very high energies. Conversely, for quasars U_{ph} can be ~ 1 and Compton losses may be the most important energy loss mechanism for relativistic electrons.

It is important to emphasize that the formulas derived here apply only in frequency range where the Thomson cross section can be used, i.e., when

$$h\nu_i' \sim \gamma h\nu_i \ll m_e c^2 \quad \text{(Thomson)} \tag{4-46}$$

In the other extreme, when

$$h\nu_i \gamma \gg m_e c^2 \quad \text{(Klein-Nishina)} \tag{4-46$'$}$$

the Klein-Nishina formula applies (see Section 4.2) and photons of energy $h\nu_f \sim m_e c^2$ are obtained in each collision. The Compton losses then increase only logarithmically with energy.

In order to calculate the spectral power $P(\nu)$ due to Thomson scattering, one can transform the ambient radiation field into the electron's rest frame, calculate $P(\nu)$ there, and then transform back into the laboratory frame. For an electron with $\gamma \gg 1$ scattering off an isotropic radiation field the result is (Jones, 1968, Blumenthal and Gould, 1970)

$$P(\nu) = 8\pi\ r_o^2 c\ h \int f(\nu/4\gamma^2 \nu_i) n_{ph}(\nu_i) d\nu_i \quad (4\text{-}47)$$

where

$$f(x) = x + 2x^2 \log x + x^2 = 2x^3 ; \quad 0 < x < 1$$

$$= 0 \qquad x > 1$$

$$(4\text{-}48)$$

and $n_{ph}(\nu_i)$ is the spectral distribution for the number of photons per unit volume. The function f(x) represents the Thomson spectrum for an electron scattering off a monochromatic radiation field, with the emitted frequency expressed in terms of the maximum Compton scattered frequency $4\gamma^2\nu_i$. Figure 4.3 is a plot of f(x), which exhibits a maximum at x = 0.61 where f = 0.16. Below the maximum frequency, for $\nu << \gamma^2\ \nu_i$, the spectrum $P(\nu) \propto \nu$ so that in the absence of absorption or small scattering angle effects the Compton spectrum cannot be cut off at low frequencies more sharply than $P(\nu) \propto \nu$.

The Compton spectrum (4-47) represents the spectrum of radiation when a single electron of energy $\gamma\ m_e c^2$ Thomson scatters with an isotropic radiation field with differential number density $n_{ph}(\nu)$. The total spectral Compton power per unit volume is obtained when this result is integrated over an electron distribution function $N(\gamma)$:

$$j(\nu) = \int P(\nu) N(\gamma) d\gamma = 8\pi\ r_o^2 hG$$

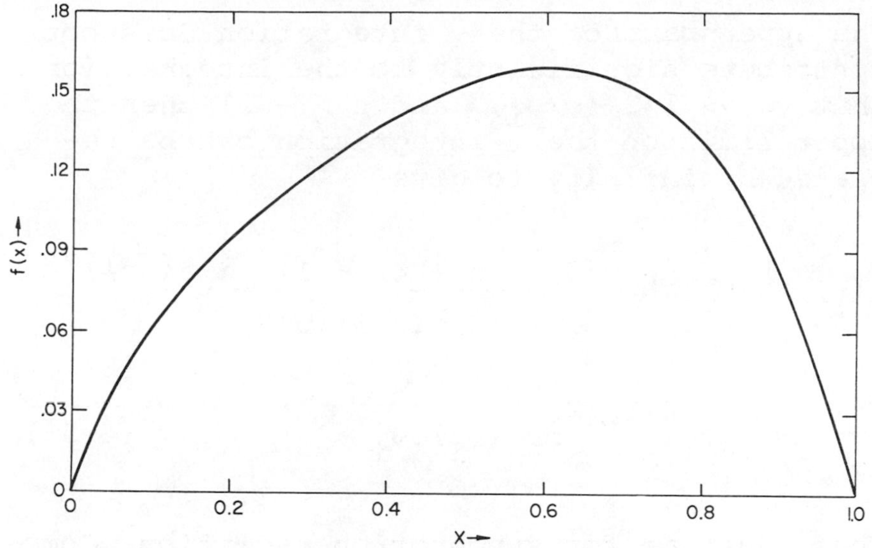

Figure 4.2. Plot of the function f(x) which represents the dimensionless Compton scattered power spectrum for a single electron.

$$G = \iint N(\gamma) f(\nu/4\gamma^2 \nu_i) n_{ph}(\nu_i) d\nu_i d\gamma \qquad (4\text{-}49)$$

As in the case of synchrotron radiation, one often deals with a power law electron distribution functions of the form

$$N(\gamma) = N\gamma^{-n} \qquad (4\text{-}50)$$

for $\gamma_1 < \gamma < \gamma_2$. Then, the lower limit in the γ-integration in Equation (4-49) is given by the larger of γ_1 and $(1/2)(\nu/\nu_i)^{\frac{1}{2}}$. We assume that we are far from the low energy end of the Compton distribution, so that $\gamma_1 < (1/2)(\nu/\nu_i)^{\frac{1}{2}}$. If we further assume that the upper part of the γ integration does not contribute significantly to the integral (or that $\gamma_2 >> 1/2\ (\nu/\nu_i)^{\frac{1}{2}}$ and $n > -1$) then the upper limit on the γ-integration can be replaced by infinity to give

$$j(\nu) = \pi r_o^2 chN2^{n+3} \frac{n^2 + 4n + 11}{(n+3)^2 (n+1)(n+5)} \nu^{-(n-1)/2}$$

$$\cdot \int \nu_i^{(n-1)/2} n_{ph}(\nu_i) d\nu_i \qquad (4\text{-}51)$$

Thus, just as for synchrotron radiation, Compton scattering of electrons with $N(\gamma) \propto \gamma^{-n}$ gives rise to a power law in ν with energy spectral index $(n-1)/2$. When the initial radiation is given by the blackbody distribution, Equation (2-21) then

$$j(\nu) = \frac{2r_o^2}{\hbar^2 c^2} b(n) N (KT)^3 (KT/h\nu)^{(n-1)/2}$$

$$= 4.2\times10^{-40}\ Nb(n) T^3 (2.1\times10^{10} T/\nu)^{(n-1)/2}$$

$$\text{erg/cm}^3\text{-sec-Hz} \qquad (4\text{-}52)$$

where

$$b(n) = \frac{2^{n+3} (n^2 + 4n + 11) \Gamma[(n+5)/2] \zeta[n+5)/2]}{(n+3)^2 (n+1)(n+5)} \qquad (4\text{-}53)$$

where ζ is the Riemann ζ-function.

The function b(n) is tabulated in Table 4.1. The assumption that the endpoints of the electron distribution do not contribute becomes $\gamma_1^2 KT/h << \nu << \gamma_2^2 KT/h$ while the Thomson limit criterion requires $(h\nu/KT)^{\frac{1}{2}} << m_e c^2$. This result was first derived by Ginzburg and Syrovatskii (1964). It may have application to the isotropic X-ray background, which can be characterized by a power law over a range frequencies.

From Equations (3-50) and (4-52) it follows that in a source with a magnetic field B, spectral index q and universal background radiation field described by a temperature T, the ratio of the intensity coefficient of the Compton spectrum k_c to that of the synchrotron spectrum k_s is given by

$$k_c/k_s = 2.47\times10^{-19}(5.25\times10^3)^q \quad T^{3+q}$$

$$\cdot\ B^{-(q+1)}\ [b(n)/a(n)] \qquad (4\text{-}53)$$

Table 4.1. The Function b(n) in Equation (4-53) from Blumenthal and Gould (1970)

n	0	1.0	1.5	2.0	2.5	3.0	4.0	5.0
b(n)	3.48	3.20	3.91	5.25	7.57	11.54	30.62	92.90

The ratio k_c/k_s is related to observed quantities by

$$k_c/k_s = (F_c/F_s)(\nu_c/\nu_s)^q \tag{4-54}$$

where F_c is the observed synchrotron flux density at frequency ν_s and F_c is the Compton flux density at frequency ν_c. Equation (4-53) is potentially of great importance to radio and X-ray astronomy since it provides us with a means of determining the magnetic field in extended radio sources by measuring the synchrotron radio emission and Compton scattered X-ray emission (see Problem 4.6).

Another case of interest occurs when the ambient radiation field is a power law of the form

$$n(\nu) = n_0 \nu_i^{-p} \tag{4-55}$$

for $\nu_a < \nu_i < \nu_b$. Then, using Equation (4-48) one can perform the ν_i integration in Equation (4-49) first. In the frequency range where $4\gamma^2\nu_a < \nu << 4\gamma^2\nu_b$, the limits on the ν_i integration may be replaced by $\nu/4\gamma^2$ and infinity to give

$$j(\nu) = 2^{2p+1} \pi r_0^2 ch\, n_0\, \nu^{-p+1}[\frac{1}{p} - \frac{2}{p+2} + \frac{1}{p+2} \cdot [\frac{1}{p} - \frac{2}{p+2} + \frac{1}{p+2} - \frac{2}{(p+1)^2}]\int d\gamma\, \gamma^{2p-2} N(\gamma) \tag{4-56}$$

Thus, if $N(\gamma)$ is a sufficiently narrow distribution that $(\gamma_2/\gamma_1)^2 << \nu_b/\nu_a$, then over the range $4\gamma_2^2 \nu_a << \nu << 4\gamma_1^2 \nu_b$, the γ-integration in Equation (4-58) becomes a constant, and the Compton scattered spectrum is a power law with the same spectral index as the initial radiation.

4.5. Compton Energy Exchange In A Hot Plasma

The energy exchange in Compton scattering of non-relativistic electrons, though small in an individual scattering, can lead to a substantial energy exchange between the electron and photon gas. Consider a monochromatic beam of photons moving through a cold plasma of temperature T. The average energy transferred per scattering from the photons to the electrons is obtained by averaging Equation (4-17) over all angles.

$$(\overline{\Delta W_e})_c = (h\nu/m_e c^2)h\nu \quad ; \quad KT << h\nu \qquad (4\text{-}57)$$

The total energy exchange rate per unit volume for a photon gas is then

$$dU_{ph}/dt = \int (\Delta W_e)_c \; N_e \; c\sigma_T(1+\overline{n}) N_{ph}(\nu) d\nu \qquad (4\text{-}58)$$

where $\overline{n}$ is the distribution function of the number of photons per unit volume of phase space and the factor $(1+\overline{n})$ takes into account stimulated as well as spontaneous scattering (see Equations (2-7) and (2- 8)). For a blackbody photon gas at a temperature T_{ph}

$$\overline{n} = 2/(e^{h\nu/KT_{ph}} -1) \qquad (4\text{-}59)$$

and $N_{ph}(\nu) = U_{ph}(\nu)/h\nu$ is given by Equation

(2-21). Evaluation of the integral yields

$$dU_{ph}/dt = -4(\sigma_T/m_e c)N_e\ KT_{ph}U_{ph},\ \text{erg/cm}^3\text{ sec} \quad (4\text{-}60)$$

with U_{ph} given by Equation (2-29).

The effect of stimulated scattering is small in this case; because of the energy dependence of the average energy exchange per scattering and in the photon distribution function, large values of $h\nu$ are weighted in the integral (4-58) so that $\bar{n}$ is small over the range of interest.

The formula for the Compton scattering energy exchange rate for finite electron temperatures can be derived from the low temperature limit (4-60). The general expression for (dU_{ph}/dt) must in general be proportional to $N_e U_{ph}$. Furthermore, it must approach (4-60) in the limit $T_e \rightarrow 0$, whereas in the limit $T_e \rightarrow T_{ph}$, (dU_{ph}/dt) must go to zero, corresponding to thermal equilibrium. The simplest functional form of T_e and T_{ph} that satisfies these conditions is $f = T_{ph} - T_e$, so that

$$(dU_{ph}/dt) = -(dU_e/dt)$$

$$= -4(\sigma_T/m_e c)N_e\ U_{ph}k(T_{ph} - T_e)$$

$$\text{erg/cm}^3\text{ sec} \quad (4\text{-}61)$$

In order to evaluate the change in the spectrum or the photon gas due to the Compton scattering, one has to use a kinetic equation for the photon distribution function (see

Gould, 1972). However, the general features can be understood by considering the escape of individual photons from the source. If the scattering cross section σ_{sc} is larger than that for absorption, σ_{abs}, then the radiation must diffuse out of the source and the effective cross section for absorption becomes (see Felten and Rees, 1971)

$$\sigma_{eff} \sim (\sigma_{sc}\,\sigma_{abs})^{1/2} \tag{4-62}$$

Furthermore, from (4-38) it follows that a low frequency photon scattering off a non-relativistic electron with velocity v experiences a frequency change $\Delta\nu/\nu \sim v/c$. For radiation scattering off an isotropic Maxwellian distribution of electrons, the lowest order term vanishes and on the average $\Delta\nu/\nu \sim (v/c)^2\ KT/m_e c^2$. Thus when the photon scatters more than mc^2/KT times before leaving the source, corresponding to an electron scattering optical depth $\tau_{sc} \sim (m_e c^2/KT)^{1/2}$, both the spectrum and total luminosity of the source are significantly changed by Compton scattering, or "Comptonization" (see Felten and Rees, 1972, and Illarionov and Sunyaev, 1972).

Finally, Compton scattering can decrease the strengths of edges and lines emitted from a hot plasma. For line radiation Compton scattering can be viewed as a random walk in frequency, so that in escaping from a source a line is broadened to

$$\Delta\nu/\nu \simeq n^{1/2}\ KT/m_e c^2 \sim \tau_{sc} KT/m_e c^2 \tag{4-63}$$

which can be sufficiently large to make

detection impossible (see Felten, Adams and Rees, 1972, and references cited therein).

4.6. Pair Annihilation And Creation

Two-Photon Pair Annihilation

From the quantum electrodynamical point of view, another phenomenon completely analogous to Compton scattering is two-photon pair annihilation. Two photons are necessary to maintain conservation of momentum and energy when pair annihilation takes place in the absence of an external potential. The cross section for the annihilation of a positron with the energy $m_e c^2 \gamma$ with an electron at rest is (Heitler, 1954, Feynman, 1962).

$$\sigma_{pair} = \frac{\pi r_o^2}{\gamma+1}\left[\frac{\gamma^2+4\gamma+1}{\gamma^2-1}\log(\gamma+(\gamma^2-1)^{1/2}) - \frac{\gamma+3}{(\gamma^2-1)^{1/2}}\right] \tag{4-64}$$

The cross section σ_{pair} has its maximum for small energies. As $\gamma \rightarrow 1$ the cross section diverges. However, the rate of annihilation per sec in a substance with N_e electrons per unit volume approaches

$$R = N_e \sigma v_+ \longrightarrow \pi r_o^2 c N_e \quad (\gamma \rightarrow 1) \tag{4-65}$$

which is approximately the rate for Thomson scattering. For the extreme relativistic case

$$\sigma_{pair} = (\pi\ r_o^2/\gamma)\ [\log 2\gamma - 1] \tag{4-66}$$

which is analogous to the Klein-Nishina formula (4-27). The two annihilation quanta do not in general have the same frequency. In the extreme relativistic case, the quantum emitted in the forward direction takes up nearly all the energy of the positron, whereas the second quantum only has an energy of the order $m_e c^2$. In the non-relativistic case the two photons have an energy of the order of $m_e c^2$ each and are emitted in opposite directions.

Upon comparing (4-66) with Equation (5-9) for ionization losses for a fast positron (or electron) we see that in most cases a fast positron will first lose all its energy and then be annihilated at a rate given by (4-65).

Pair Creation By γ-rays In The Field Of A Nucleus. The energy necessary to create an electron-positron pair is $\gtrsim 2m_e c^2$. Energy and momentum conservation are possible only if another particle is present. Consider first the pair creation by γ-rays in the presence of a nucleus. There are two indistinguishable ways in which this can happen: (i) the incoming photon creates a pair and subsequently the electron interacts with the field of the nucleus, or (ii) the photon creates a pair and the positron interacts with the field of the nucleus. The cross section for pair creation in the field of a nucleus is found to be (Heitler, 1954; Feynman, 1962)

$$\sigma(W_+)dW_+ = (4\ \alpha_f r_o^2 Z^2/h\nu)(1 - (4/3)W_+W/(h\nu)^2)G$$

$$G \cdot x(\log 2W_+W_+/h\nu\, m_ec^2 - 1/2)dW_+ \quad (4\text{-}67)$$

when the positron energy W_+ and the electron energy W_- are both $\gg m_ec^2$ ($\alpha_f = e^2/\hbar c$ is the fine structure constant). The cross section is symmetrical between the positron and electron energy.

If the energy of the electrons is so high that the screening of the Coulomb field is important ($2\alpha_f Z^{1/3} W_+W_- \ll m_ec^2 h\nu$, see Chapter 5) then the argument of the logarithm becomes $183Z^{-1/3}$ and the factor 4/3 in the expression multiplying the logarithm becomes 13/9.

Note the similarity between the cross section for pair creation (4-67) and that for bremsstrahlung in the extreme relativistic case (5-72).

The total number of pairs created is obtained by integrating over all possible energies of the positron. In the extreme relativistic case, when screening is negligible

$$\sigma_{pair} = Z^2\alpha_f r_o^2\left(\frac{28}{9}\log\frac{2h\nu}{m_ec^2} - \frac{218}{27}\right) \quad (4\text{-}68)$$

For complete screening

$$\sigma_{pair} = Z^2\alpha_f r_o^2 \frac{28}{9}\left(\log\frac{183}{Z^{1/3}} - \frac{1}{42}\right) \quad (4\text{-}69)$$

Comparison with the Klein-Nishina formula (4-27) multiplied by the number of atomic electrons shows that pair creation becomes a more important cause of photon attenuation than the Compton effect when $h\nu \gg Z\, m_e c^2/\alpha_f$.

Pair creation by photons and bremsstrahlung are the causes of electron-photon cascade showers produced by cosmic radiation. For example, a photon of extremely high energy (produced for instance, by the decay of a π^0 meson) is absorbed and creates a high energy pair. They lose most of their energy by bremsstrahlung, and the high energy photons thus produced produce more pairs, etc.

The maximum number of shower particles is, to an order to magnitude given by (Heitler, 1954)

$$N_m \sim W_0/30\, W_c \tag{4-70}$$

where W_0 is the energy of the primary electron or photon and $W_c \sim 155\, m_e c^2$ for air. The shower maximum occurs at a depth

$$d_{max} \sim d_0 \log(W_0/3W_c) \tag{4-71}$$

where $d_0 \approx 3\times10^4$ cm for air.

References

4.1. Thomson Scattering

Jackson (1962) and Landau and Lifshitz (1962). For a discussion of scattering by a system of charges, see Jackson (1962). Scattering from density fluctuations in a plasma is discussed by Bekefi (1966).

4.2. Compton Scattering and Klein-Nishina Formula

Feynman (1962) and Heitler (1954)

4.3. Radiation Pressure
Landau and Lifshitz (1962) and
Eddington (1926)

4.4. Scattering by Moving Electrons
Ginzburg and Syrovatskii (1964)

Blumenthal and Gould (1970)

Felten and Morrison (1966)

Jones (1968)

Woltjer (1966)

4.5. Compton Energy Exchange in a Hot Plasma
Gould (1972)

Weymann (1965)

Felten and Rees (1972)

Illarionov and Sunyaev (1972)

Gnedin and Sunyaev (1973)

4.6. Pair Annihilation and Creation
Feynman (1962)

Heitler (1954)

4.7. Other Topics
Electron scattering in a strong magnetic field

Canuto, Lodenquai and Ruderman (1971)

Problems

4.1. Compute the optical depth for Thomson scattering of

(a) The Sun, assuming that it is a sphere of fully ionized hydrogen gas with a uniform density having a mass of 2×10^{33} gm and a radius of 7×10^{10} cm.

(b) The disk of our galaxy, assuming an average electron density of 0.1 cm^{-3}.

(c) Intergalactic space out to the Hubble radius assuming an electron density of 10^{-5} cm^{-3}.

4.2. Consider a quasar which is surrounded by a fully ionized hydrogen gas cloud having a density $N_e = 10^6$ cm^{-3} and a radius of 10^{20} cm. If the quasar undergoes a change of intensity at the source a time scale t_1, estimate the time scale for the observed variation in intensity.

4.3. Compute the average energy loss for 1 MeV gamma rays scattering off electrons at rest.

4.4. Consider a crystal of cesium iodide having an average atomic number Z = 50, atomic weight A = 130 and density $\zeta = 4.5$ gm/cm^3. What thickness of the crystal is needed to scatter 95% of a normally incident flux of 1 MeV gamma rays?

Compute the mean free paths for Compton scattering and absorption of 10 MeV gamma rays in this crystal.

4.5. Centaurus A is an extended source of synchrotron radio emission. The 1-100 MHz radio data can be fit to a curve of the form

$$F_r(\nu) = k_r \nu^{-0.9}$$

with

$$k_r = 1\times10^{-12}.$$

The upper limit on the 1-10 keV X-ray emission from the extended source imply that the intensity coefficient k_c of the Compton spectrum produced by scattering the microwave background radiation off the synchrotron electrons is

$$k_c < 7\times10^{-13}.$$

Using a temperature $T = 2.7^oK$, find the lower limit on the magnetic field in the extended source. How does it compare to the equipartition field (see Table 3.2)?

For a more difficult problem of this type, see J. Grindlay and J. Hoffman, Astro. Letters 8, 209 (1971).)

4.6. Evaluate the Compton energy loss rate for the hot electrons in (a) the solar corona ($N_e = 10^9$, $T_e = 10^6$ ^oK, $T_{ph} = 6000^oK$) and (b) a compact X-ray source having a luminosity of 10^{38} erg/sec which is surrounded by a cloud of cold electrons having a density of 10^{15}

cm^{-3} and a radius of 10^8 cm.

4.7. Consider a source of radius R with a synchrotron luminosity L frequency. Compute the magnetic field needed if synchrotron losses are to exceed the Compton losses. What is the lower limit on the optical depth for synchrotron reabsorption in these sources?

4.8. Compute the density and spectrum of cosmic ray electrons in intergalactic space needed to produce the observed X-ray background spectrum of

$$I(h\nu) = 40(h\nu)^{-1.2} \text{ keV } (\text{cm}^2 \text{ sec ster keV})^{-1}$$

in the range 20 keV to 1 MeV. Neglect cosmological effects and integrate out to a Hubble radius $\simeq 2\times10^{28}$ cm.

Chapter 5

BREMSSTRAHLUNG AND COLLISION LOSSES

Whenever an electron traverses an ionized gas it may lose energy by (a) non-radiative collisions in which the electron excites an atom or ion to a higher state or ionizes it, or excites plasma oscillations in a fully ionized plasma; (b) radiative bremsstrahlung or free-free collisions in which the electron makes a transition from one free state to another while emitting a photon; or (c) recombining with an ion an emitting recombination or free-bound radiation as it makes the transition from a free to a bound state.

Non-radiative collisional losses and bremsstrahlung are considered in this chapter. Radiative recombination is discussed in Chapter 6. First we consider the non-radiative collisions.

5.1. Collisional Or Ionization Losses

If we denote the distance of closest approach or impact parameter by b, then the momentum impulse Δp which the target electron receives is of the order

$$\Delta p \simeq \int_0^{\tau} F\, dt \simeq (e^2/b^2)(2b/v) = 2\, e^2/bv \quad (5\text{-}1)$$

where F is the Coulomb force and $\tau \simeq 2b/v$ is the collision time. The energy transferred is therefore

$$\Delta W(b) \simeq (\Delta p)^2/2m \simeq 2\, e^4/mv^2 b^2 \quad (5\text{-}2)$$

An electron passing through a medium will

encounter target electrons at various distances or impact parameters. If there are N_e target electrons per unit volume, then an electron moving through this medium with a velocity v will encounter $2\pi bv\, db$ targets per second. The energy loss rate for such an electron is therefore

$$dW/dt = 2\pi N_e v \int \Delta W(b) b \, db \qquad (5\text{-}3)$$

We can obtain an order of magnitude estimate of the collisional energy loss rate by substituting (5-2) into (5-3) and integrating over all impact parameters. The result is

$$dW/dt \simeq (4\pi N_e e^4 / m_e v) \log(b_{max}/b_{min}) \qquad (5\text{-}4)$$

where the maximum and minimum impact parameters, b_{max} and b_{min}, are determined by the conditions of the problem. The maximum impact parameter is given by the requirement that the collision time be short compared with the characteristic period $1/\nu_o$ of motion of the gas, otherwise the electron will interact adiabatically with the Coulomb fields of the target electrons. Thus

$$b_{max} \simeq v/\nu_o \qquad (5\text{-}5)$$

For a gas composed mainly of atoms,

$$\nu_o = I/h \qquad (5\text{-}6)$$

where I is the average excitation energy which is slightly less than the ionization energy. In this case the energy of the electrons is used to excite the gas. For a highly ionized gas ν_o is equal to the plasma frequency and the energy of the electrons goes into the excitation of plasma oscillations. In either case most of the energy ultimately goes into heating the gas, so non-radiative collisional losses are often referred to as heating losses.

For relatively slow moving incident electrons, the minimum impact parameter is given by the requirement that the maximum energy transfer in classical collisions is $m_e v^2/2$, so

$$b_{min} = 2\, e^2/m_e v^2 \qquad (v < \alpha_f c) \qquad (5\text{-}7)$$

where the condition in parentheses gives the definition of slow incident electrons. The origin of this definition is apparent when we consider the limitations imposed by the uncertainty principle. It requires that the quantum mechanical uncertainty in the momentum of the incoming electron be less than its classical momentum $m_e v$. The minimum impact parameter defined in this way is

$$b_{min} = \hbar/m_e v \qquad (v > \alpha_f c) \qquad (5\text{-}8)$$

From Equations (5-7) and (5-8) we see that the quantum mechanical limitation on the impact parameter is the most important one for electron velocities $v > \alpha_f c$ where α_f is the fine structure constant.

An exact calculation verifies that the order of magnitude estimates are fairly accurate. The exact results are (Gould, 1972 a, b)

$$dW/dt = (4\pi \; N_e e^4 / m_e v) B \tag{5-9}$$

where, for atomic matter, $N_e = ZN_{atom}$ and

$$B = \log(W_e/2^{\frac{1}{2}}I) + 1/2 - v^2/2c^2 \qquad (v > \alpha_f c) \tag{5-10}$$

For an ionized gas,

$$B = \log(W_e/2^{\frac{1}{2}} \, h\nu_p) + 1/2 \qquad (v > \alpha_f c) \tag{5-11}$$

where in both cases W_e is the kinetic energy of the electron and $v \ll c$.

In the extreme relativistic case (Gould, 1972 a,b)

$$B = \log(\gamma^{3/2} \; m_e c^2 / 2^{\frac{1}{2}} I) + (1/16) \tag{5-12}$$

for atomic matter, and

$$B = \log(\gamma^{\frac{1}{2}} \; m_e c^2 / 2^{\frac{1}{2}} h\nu_p) + (9/16) \tag{5-13}$$

for a fully ionized gas.

5.2. Bremsstrahlung-Classical Calculation

Bremsstrahlung radiation is emitted when an electron is accelerated in the Coulomb field on an ion (see Figure 5.1). The massive target particle will be accelerated only very slightly and will radiate a negligible amount compared to the electron, so it is sufficient to treat the collision as the interaction with a fixed field of force. Strictly speaking, bremsstrahlung is a quantum process since

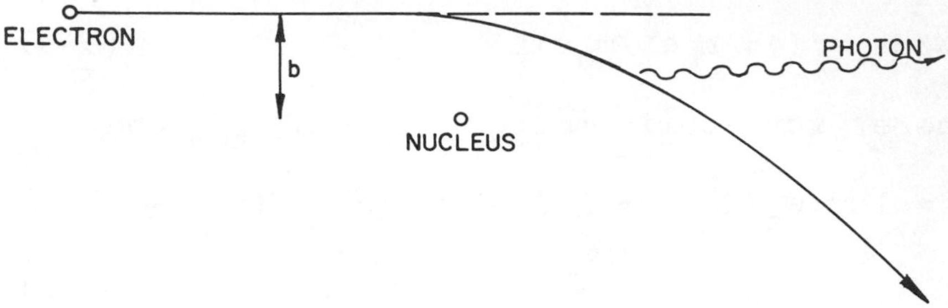

Figure 5.1. Bremsstrahlung emission associated with the collision of an electron with a nucleus.

since most of the energy is radiated in photons having energies of the same order as the electron energy. However, the classical calculation yields results which are quite close to the correct quantum results.

For a non-relativistic electron velocities the energy radiated throughout the collision as given by the classical Larmor formula is (Equation (1-147))

$$W(\omega,b) = 8\pi(\ddot{d}(\omega))^2/3c^3 \qquad (5\text{-}14)$$

Note that a system consisting of two electrons or two protons cannot emit dipole radiation, since in these cases the second time derivative of the dipole moment vanishes. Thus only electron-ion collisions will produce non-relativistic bremsstrahlung radiation. Since the ion is much more massive than the electron, we can assume to a good approximation that the center of mass of the electron-ion system is at the position of the ion. Then the dipole moment of the system is

$$\vec{d} = e\vec{r} \tag{5-15}$$

and

$$\ddot{\vec{d}}(\omega) = (e/2\pi) \int_{-\infty}^{\infty} \dot{\vec{v}}\, e^{i\omega t}\, dt \tag{5-16}$$

To compute the Fourier components we need the equation of motion for the electron. For an electron having a velocity v relative to the ion at infinity, and an impact parameter b (see Figure 5.1), the motion of the electron can be written in the parametric form

$$x = b_o(\epsilon - \cosh\xi); \quad y = b \sinh\xi ;$$

$$t = (b_o/v)(\epsilon \sinh\xi - \xi) \tag{5-17}$$

where the parameter ξ runs from $-\infty$ to $+\infty$, and

$$\epsilon = (1 + (b/b_o)^2)^{\frac{1}{2}} \quad ; \quad b_o = Ze^2/m_e v^2 \tag{5-18}$$

Note that b_o is the impact parameter corresponding to a scattering angle of 90 degrees.

Using Equations (5-17) and (5-18), the Fourier components can be computed in terms of Hankel function of the first kind and order $i\omega b_o/v$ (see Landau and Lifshitz, 1962). To find the energy radiated per second by an electron passing through a plasma having a density N_Z per cubic centimeter of ions with charge Z, an integration over impact parameters must be performed (see Equation (5-3)):

$$P(\omega) = 2\pi\ N_Z v \int_{-\infty}^{\infty} W(\omega, b) b\ db$$

$$= (4\pi^2/3)\ v^2 N_Z (e^2/c^3)\omega\ b_o^3\ \Xi \qquad (5\text{-}19)$$

where

$$\Xi = \left| H^{(1)}_{i\omega b_o/v}(i\omega b_o/v) \right| H^{(1)\prime}_{i\omega b_o/v}(i\omega b_o/v) \qquad (5\text{-}20)$$

In Equation (5-20) the H's are Hankel functions and the prime denotes differentiation with respect to the argument.

For low frequencies, $\omega \ll v/b_o$, corresponding to small angle scattering,

$$\Xi \cong (4v/b_o \omega\pi^2)\log(2v/\Gamma\omega b_o)\ ;\ \omega \ll v/b_o \qquad (5\text{-}21)$$

where $\Gamma = 1.78$. The spectral power at low frequencies is

$$P(\nu) = 2\pi P(\omega) = (32\pi/3)\ v^3 N_Z (e^2/c^3)\ b_o^2 \cdot \log(2v/\Gamma\omega b_o) \qquad (5\text{-}22)$$

The effective differential cross section for the emission of a photon with energy between $h\nu$ and $h\nu + dh\nu$ is defined as

$$\sigma(\nu) = P(\nu)/N_Z v\ h\nu \qquad (5\text{-}23)$$

In the case of low frequency classical bremsstrahlung,

$$\nu\sigma(\nu) = (16/3)(v/c)^2\ \alpha_f\ b_o^2\ \log(2v/\Gamma\omega b_o)$$

$$= (16/3)\ \alpha_f^3\ k_e^{-2}\ Z^2\ \log(\ldots)\ ;$$

$$\omega << m_e v^3/Ze^2 \qquad (5\text{-}24)$$

where

$$k_e = m_e v/\hbar \qquad (5\text{-}25)$$

k_e^{-1} is the De Broglie wavelength of the electron.

In the limit of high frequencies, $\omega >> v/b_o$ corresponding to large angle scattering,

$$\Xi \simeq (4/\pi 3^{1/2})(v/\omega b_o) \quad ; \quad \omega << v/b_o \qquad (5\text{-}26)$$

and

$$\nu\sigma(\nu) = (16\pi/3^{3/2})\ (v/c)^2\ \alpha_f\ b_o^2$$

$$= (16\pi/3^{3/2})\ \alpha_f^3\ k_e^{-2}\ Z^2\ ;$$

$$\equiv \sigma_o \qquad \omega >> m_e v^3/Ze^2 \qquad (5\text{-}27)$$

where we have given the right hand side of (5-27) the special name σ_o because the bremsstrahlung cross section is usually expressed in terms of the classical high frequency cross section. Thus it is customary to write the bremsstrahlung cross section as

$$\nu\sigma(\nu) = \sigma_o\ g_{ff}(\nu, v) \qquad (5\text{-}28)$$

where g_{ff} is the free-free or bremsstrahlung Gaunt factor which is in general a function of frequency and velocity.

From Equations (5-24) and (5-27) we see that the Gaunt factors for classical bremsstrahlung are

$$g_{ff}(\nu, v) = (3^{\frac{1}{2}}/\pi) \log(m_e v^3/5.6\, Ze^2 \nu) \ ; \qquad \nu \ll m_e v^3/2\pi Ze^2$$

$$= 1 \ ; \qquad \nu \gg m_e v^3/2\pi Ze^2 \qquad (5\text{-}29)$$

These expressions were derived using the classical dipole approximation. The dipole approximation is valid only for electron velocities $v \ll c$, and the classical description valid only for frequencies $\nu \ll m_e v^2/2h$. Combining this last inequality with the lower limit on the frequency as given by the second half of Equation (5-29), we see that the classical result is valid only for velocities $v \ll Z\alpha_f c$. A similar limit can be obtained for the low frequency classical bremsstrahlung by considering the limitations on the minimum impact parameter which are imposed by the uncertainty principle (cf. Equation (5-7)). Thus, for large electron velocities and high frequencies, the quantum mechanical results must be used. Before considering these, an order of magnitude derivation will be given of Equations (5-24) and (5-27), or equivalently (5-29).

5.3. Classical Bremsstrahlung-Order of Magnitude Estimates

To obtain an order of magnitude estimate of

the bremsstrahlung radiation in the classical dipole approximation, we neglect the exponential term in the Fourier integral (5-16). This is justified since the acceleration $\dot{\vec{v}}$ is negligible except for a time τ, where τ is the duration of the collision. For high frequencies $\omega\tau >> 1$ the integral oscillates rapidly and the contributions of the integrand interfere destructively. For low frequencies such that $\omega\tau << 1$, we have

$$\ddot{\vec{d}}(\omega) \simeq (e/2\pi) \int_0^\tau \dot{\vec{v}}\, e^{i\omega t} dt \simeq (e/2\pi) \int_0^\tau \dot{\vec{v}} dt$$

$$\simeq (e\, \Delta\vec{v}/2\pi) \qquad (5\text{-}30)$$

where $\Delta\vec{v}$ is the change in the electron's velocity during the collision.

The radiation emitted during an electron-ion collision is therefore on the order of

$$W(\omega, b) = (2/3\pi)(e^2/c^3)(\Delta v)^2 \qquad (5\text{-}31)$$

For small angle collisions (distant encounters) Δv is given by

$$m_e \Delta v \simeq \int_0^\tau F\, dt \simeq (2\, Ze^2/bv) \qquad (5\text{-}32)$$

Substituting Equation (5-32) into (5-31) and integrating over impact parameters yields an expression for the Gaunt factor:

$$g_{ff} = (1/\sigma_o \hbar) \int_{b_{min}}^{b_{max}} W(\omega, b)\, 2\pi b\, db$$

$$= (3^{1/2}/\pi) \log(b_{max}/b_{min}) \qquad (5\text{-}33)$$

The minimum impact parameter can be estimated from the assumption that the electron is scattered through a small angle. If we say that all deflections for which the scattering angle $\Delta\theta \simeq \Delta v/v$ is less than unity are small angle collisons, then

$$b_{min} = 2Ze^2/mv^2 = 2b_o \tag{5-34}$$

The maximum impact parameter is given by the condition that $\omega\tau \ll 1$, so

$$b_{max} = 2v/\omega \tag{5-35}$$

and the Gaunt factor $g_{ff} \simeq (3^{\frac{1}{2}}/\pi)\log(mv^3/Ze^2\omega)$, which is identical to the low frequency limit given by the exact calculation except that the factor 5.6 in the logarithm is replaced by 2π.

For large angle scattering $\Delta v \simeq 2v$, so $W(\omega,b) \simeq (2/3\,\pi)(e^2/c^3)(4v^2)$ and

$$g_{ff} \simeq (1/\sigma_o\hbar)\int_{b_{min}}^{b_{max}} W(\omega,b)\,2\pi b\,db$$

$$= (3^{\frac{1}{2}}/2\pi)(b_{max}/b_o)^2 \tag{5-36}$$

Choosing the maximum impact parameter for large angle scatterings as that impact parameter which produces a scattering angle $\Delta\theta \simeq 1$ namely, $2b_o$, we obtain $g_{ff} = (3^{\frac{1}{2}}/2\pi)$ which is about ten percent higher than the exact high frequency result.

5.4. Bremsstrahlung--Non-relativistic Born Approximation

Consider now the quantum mechanical treatment of bremsstrahlung. The transition probability per second for the emission of a bremsstrahlung photon of frequency $h\nu$ into the solid angle $d\Omega$ in the interaction of an electron with a nucleus of charge Z is (see Equations (2-118) and (2-135))

$$dw(\nu)/d\Omega = (e^2\nu 2\pi/m_e^2c^3h)\left|(e^{-i\vec{k}\cdot\vec{r}}\ \vec{1}\cdot\vec{p})_{fi}\right|^2 \cdot (dn/dW) \quad (5\text{-}37)$$

The cross section is given by

$$d\sigma(\nu)/d\Omega = (dw(\nu)/d\Omega)/v \quad (5\text{-}38)$$

It is assumed that the nucleus is infinitely heavy, so that it can absorb momentum but not energy.

The wave functions for electrons in a Coulomb potential field are known so the bremsstrahlung cross section can be calculated exactly in terms of hypergeometric functions (Sommerfeld, 1939). Before discussing the general properties of this exact solution, the Born approximation result will be discussed. In this approximation the potential due to the ion is considered as a small perturbation and a single plane wave with a specific momentum is used as the unperturbed wave function.

The restriction imposed on the electron velocity can be estimated by considering that the minimum impact parameter must be on the order of the spread in the wave packet of the electron as given by the uncertainty principle

$$b_{min} \simeq k_e^{-1} = \hbar/m_e v \qquad (5\text{-}39)$$

At this distance the effect of the Coulomb potential will be greatest. We have

$$(Ze^2/b_{min})/W_i \simeq (Ze^2 m_e v/\hbar)/(m_e v^2/2)$$

$$= 2Z\alpha_f\ c/v \qquad (5\text{-}40)$$

where W_i is the kinetic energy of the incident electron. Therefore the Born approximation is valid for electron velocities sufficiently large that

$$v \gg Z\ \alpha_f\ c \qquad (5\text{-}41)$$

In order that the final kinetic energy of the electron be much greater than its potential energy, the energy of the emitted photon must satisfy the inequality

$$h\nu \ll W_i - 2Z^2\ \alpha_f^2\ m_e c^2 \qquad (5\text{-}42)$$

The Born approximation thus gives us the low frequency spectrum for high incident electron velocities. It complements the expression (5-24) which is for low electron velocities.

In the non-relativistic Born approximation, which is valid for initial electron energies W_i much less than $m_e c^2$ and ionic charge Z much less than 137, the initial and final states of the electron can be treated non-relativistically. In this case Schrödinger wave functions can be used and

$$h\nu = W_i - W_f = (p_i - p_f)(v_f + v_i)/2 \tag{5-43}$$

The unperturbed wave functions are

$$u_{io} = (2\pi)^{-3/2}\, e^{i\vec{k}_i\cdot\vec{r}} \; ; \; u_{fo} = (2\pi)^{-3/2}\, e^{i\vec{k}_f\cdot\vec{r}} \tag{5-44}$$

where $k_i = p_i/\hbar$, etc. Using first order perturbation theory the first order wave function is found to be

$$u_{i1} = \int (2\pi)^{-3/2} \int \left\{\frac{(Ze^2/r)_{pi}}{W_i - W_p}\right\} e^{ip\cdot r/\hbar}\, d^3p \tag{5-45}$$

where

$$(Ze^2/r)_{pi} = \int (2\pi)^{-3} (Ze^2/r) e^{-i\vec{p}\cdot\vec{r}/\hbar}\, e^{i\vec{k}_i\cdot\vec{r}}\, dV \tag{5-46}$$

Upon substituting Equations (5-44) and (5-45) into the matrix element in Equation (5-37) we obtain a sum of four integrals. The leading term is proportional to $\delta(\vec{k}_i - \vec{k}_f - \vec{k})$. However, $\vec{k}_i - \vec{k}_f - \vec{k}$ can never vanish since energy must be conserved in the emission of the photon. Therefore the leading term is identically zero. For bremsstrahlung to occur the Coulomb potential must absorb some momentum. Therefore, we have to consider the effect of the potential on at least one of the wave functions u_i and u_f.

The interaction causing the transition consists of two parts: (i) the interaction of the electron with the radiation field giving rise to the emission of a photon and (ii) the interaction of the electron with the potential of the atom or ion. There are two indistinguishable orders in which these processes can occur: (a) the electron interacts with the Coulomb potential and subsequently emits a photon or (b) the electron first emits a photon and then interacts with the Coulomb potential. These two processes correspond to the two terms

$$\int u_{fo}\, e^{-i\vec{k}\cdot\vec{r}}\, \hat{\ell}\cdot\vec{p}\, u_{i1}\, dV \tag{5-47}$$

and

$$\int u_{f1}\, e^{-i\vec{k}\cdot\vec{r}}\, \hat{\ell}\cdot\vec{p}\, u_{io}\, dV \tag{5-48}$$

entering into the evaluation of the matrix element in Equation (5-37).

The evaluation of the matrix elements is straightforward but tedious. A considerable simplification is effected by neglecting terms of the order of $h\nu/c$ in comparison to terms of the order of $(p_f - p_i)$, which is the same as neglecting terms of the orcer of v/c (see Equation (5-43)). This is the quantum mechanical analog of the classical dipole approximation (see Equation (1-135)).

In the dipole approximation the differential cross section is found to be (see Heitler (1954), Chapter 5, Frank-Kamenetskii (1962), Chapter 6).

$$\sigma(\Omega_f, \Omega_k, \nu) = (2Z^2/\pi^2)\alpha_f^3\, q^{-2}\, \nu^{-1}\,(p_f/p_i)\cos^2\theta \tag{5-49}$$

where $\vec{q} = \vec{k}_i - \vec{k}_f$ is the wave vector of the momentum change of the electron, and θ is the angle between $\vec{q}$ and the direction of polarization of the radiation.

The cross section obtained upon integration over the direction of the emitted photon, i.e., over θ, the direction of electron after the interaction, and integrating q^{-2} over solid angle, is (see Equation (5-27))

$$\sigma(\nu) = \sigma_o\, g_{ff}/\nu \tag{5-28}$$

with

$$g_{ff}(v, \nu) = (\sqrt{3}/\pi)\log\{(v_i + v_f)/(v_i - v_f)\} \tag{5-50}$$

This expression is valid in the regime where v_i and v_f are both much greater than $Z\alpha_f c$, yet are still much less than c. In this approximation the argument of the logarithm is approximately equal to $2W_i/h\nu$ for low photon energies. Thus the low frequency quantum mechanical and classical results differ only by a logarithmic factor which is essentially due to taking $b_{min} = \hbar/m_e v$ rather than $b_{min} = Ze^2/m_e v^2$. In the low frequency limit the Born approximation gaunt factor is greater than unity. It slowly decreases with increasing frequency or photon energy until at $h\nu = W_i/2$, $g_{ff} = 0.97$.

For low initial electron velocities, or

for photon energies close to W_i, corresponding to small final electron velocities, we must use an approximation based on the exact Coulomb wavefunctions for the problem.

5.5. Bremsstrahlung--Exact Results For The Non-relativistic Dipole Approximation

Sommerfeld (1939) has obtained an expression for the cross section integrated over all angles in terms of the general hypergeometric functions for an unscreened Coulomb potential in the non-relativistic case. The result is (see Equation (5-27)).

$$\sigma(\nu) = \sigma_o \, g_{ff}/\nu \tag{5-28}$$

with

$$g_{ff} = \frac{3^{\frac{1}{2}}\pi \, x_o (d|F(-in_i, -in_f, 1, x_o)|^2/dx)}{(\exp(2\pi n_i)-1)(1-\exp(-2\pi n_f))} \tag{5-51}$$

where

$$n_i = Ze^2/\hbar v_i \quad ; \quad n_f = Ze^2/\hbar v_f$$

$$x_o = -4\, v_i v_f/(v_i - v_f)^2 \tag{5-52}$$

and F(...) is the general hypergeometric function.

The classical formulas (5-24) and (5-27) can be obtained from this formula by passing to the limit of electron velocities much less than $Z\alpha_f c$, and the Born approximation from the limit of initial and final electron velocities much greater than $Z\alpha_f c$. The exact results have been tabulated by Karzas and

Latter (1961) for a wide range of electron velocities and photon energies. An interpolation formula which is fairly accurate over the whole range of electron and photon energies has been derived by Elwert (1939):

$$g_{ff} = (3^{\frac{1}{2}}/\pi)(v_i/v_f)\,\frac{(1-\exp(-2\pi n_i))}{(1-\exp(-2\pi n_f))}$$

$$\cdot\ \log\left(\frac{v_i + v_f}{v_i - v_f}\right) \qquad (5\text{-}53)$$

The values of this Gaunt factor for several cases of interest are listed in Table 5.1 The general behavior of the Gaunt factor is as follows. For low frequencies it is greater than unity, of the order of five, regardless of the initial electron velocity. For moderate frequencies, such that the final velocity of the electron is about half its initial velocity, the Gaunt factor is of the order unity. For high frequencies, such that the final velocity of the electron is approximately zero, the Gaunt factor is (a) much less than unity but finite at the hard photon limit (photon energy = initial electron velocity) for high initial electron velocities, (b) slightly less than unity for moderate initial velocities, and (c) slightly greater than unity for low initial velocities.

The energy loss per unit time due to bremsstrahlung for an electron with energy W traversing a gas with ion density N_z is

$$dW/dt = N_z v_i \int_0^{W/h} h\nu\sigma(\nu)\,d\nu$$

Table 5.1. Gaunt Factor According To Elwert's Interpolation Formula

v_i	v_f	g_{ff}
$>> 2\pi Z\alpha_f c$	$>> 2\pi Z\alpha_f c$	$(3^{1/2}/\pi)\log((v_i + v_f)/(v_i - v_f)) > 1$
$>> 2\pi Z\alpha_f c$	$\simeq 0$	$(6.9)Z\alpha_f c/v_i << 1$
$= 2\pi Z\alpha_f c$	$= v_i/2$	0.89
$= 2\pi Z\alpha_f c$	$\simeq 0$	0.69
$<< 2\pi Z\alpha_f c$	$\simeq v_i$	$(3^{1/2}/\pi)\log(m_e v^3/5.6\ Ze^2 \nu)^* > 1$
$<< 2\pi Z\alpha_f c$	$\simeq v_i/2$	1.2
$<< 2\pi Z\alpha_f c$	$\simeq 0$	1.1

* Classical limit; see Equation (5-29)

$$= N_Z v_i \sigma_o \int_o^{v_i} g_{ff}(v_f)\, m_e v_f dv_f \qquad (5\text{-}54)$$

Using the Born approximation for the cross section we find

$$dW/dt = (3.46/\pi)\, N_Z\, \sigma_o W v_i$$

$$= (16/3)\, N_Z\, Z^2 \alpha_f\, r_o^2\, m_e c^2\, v_i \qquad (5\text{-}55)$$

(r_o = classical electron radius). Note that the energy loss per unit length is independent of the electron energy W in the Born approximation. The Born approximation gives a cross section which goes to zero at the hard photon limit, so the above formula underestimates the energy loss rate. At high velocities the error introduced by the use of the Born approximation is on the order of $(2\pi\, Z\alpha_f/v_i)$, which is negligible. At low frequencies the errors become somewhat larger, on the order of 50%.

Comparing Equations (5-55) and (5-9) we see that, for non-relativistic electron energies the ratio of bremsstrahlung ionization power losses is $(N_e = ZN_Z)$

$$(dW/dt)_{ff}/(dW/dt)_{ion} = (4/3\pi)(Z\alpha_f/B)(v_i/c)^2$$

$$\simeq 3\times 10^{-3}(v_i/c)^2\,(Z/B) \qquad (5\text{-}56)$$

where B is a logarithmic factor which is on the order of ten.

Equation (5-56) shows that bremsstrahlung from a non-thermal beam of non-relativistic electrons is an inefficient radiation mechanism. In general non-thermal bremsstrahlung is a likely source of radiation from an astrophysical object only in a transient flare-like events. Indeed non-thermal bremsstrahlung is observed in solar flares where it produces the hard X-ray component (see the problems at the end of this chapter). In a hot plasma with a Maxwellian distribution of particle velocities the collisional losses only serve to transfer energy between the components of the plasma and do not represent a net energy loss. Under such conditions bremsstrahlung can be a major cooling mechanism for the plasma (see Figure 8.1).

5.6. Bremsstrahlung From A Maxwellian Plasma

For most astrophysical applications we need to know the spectrum of the bremsstrahlung radiation from an ensemble of particles. In general, the distribution of electron velocities can be assumed to be Maxwellian to a good approximation:

$$N(v)dv = 4\pi\ N_e (m_e/2\pi\ KT)^{3/2}\ e^{-m_e v^2/2KT}\ v^2 dv \qquad (5\text{-}57)$$

The emissivity (energy emitted per unit volume per second) is

$$j(\nu) = N_Z \int_{(2h\nu/m_e)^{\frac{1}{2}}}^{\infty} N(v_i) h\nu\sigma(v_i,\nu) v_i dv_i$$

$$= (64\pi/3^{3/2}) N_e N_Z Z^2 (e^2/c) r_o^2 c (c/\bar{v}_{th}) \cdot \bar{g}_{ff}(\nu,T) e^{-h\nu/KT}$$

$$= 6.8\times10^{-38}\, Z^2 N_e N_Z T^{-\frac{1}{2}}\, \overline{g}_{ff}(\nu, T) e^{-h\nu/KT}$$

$$\text{erg/cm}^3 \text{ sec Hz} \qquad (5\text{-}58)$$

where $\overline{v}_{th}$ is the average thermal velocity of the electrons of the plasma:

$$\overline{v}_{th} = (2KT/\pi m_e)^{\frac{1}{2}} \qquad (5\text{-}59)$$

and $\overline{g}_{ff}(\nu, T)$ is the temperature averaged Gaunt factor. It has been computed by Karzas and Latter (1961). From Table 5.1 we can construct the general behavior of the temperature averaged Gaunt factor shown in Table 5.2 which provides a rough summary of their calculations.

Table 5.2. The Temperature Averaged Gaunt Factor $g_{ff}(\nu, T)$

$T(^oK)$	$\nu(Hz)$	$g_{ff}(\nu, T)$
$< 3\times10^5 Z^2$	$< 10^9 T^{3/2}$	$(3^{\frac{1}{2}}/\pi)[17.7 + \log (T^{3/2}/\nu Z)]$
$< 3\times10^5 Z^2$	$> 10^9 T^{3/2}$	$\simeq 1$
$> 3\times10^5 Z^2$	$<< kT/h$	$(3^{\frac{1}{2}}/\pi)\log(2.2KT/h\nu)$
$> 3\times10^5 Z^2$	$\simeq kT/h$	$\simeq 1$
$> 3\times10^5 Z^2$	$> kT/h$	$\simeq (kT/h\nu)^{0.4}$ *

* From a numerical fit by Gorenstein *et al*., (1968).

From Equation (5-55) and Table 5.2 we see that the bremsstrahlung spectrum is flat at low frequencies, and exponential at high frequencies. Many diffuse sources of cosmic radio and microwave emission have flat continua that are almost certainly due to electron-ion bremsstrahlung from a thermal plasma (see, for example, Aller and Liller, 1968). Also there are many cosmic X-ray sources which exhibit an exponential behavior at high temperature which is well fit by a spectrum of the form (5-55) (see Giacconi and Gursky, 1974). Some selected sources whose spectra are probably due to bremsstrahlung from a hot plasma are listed in Table 5.3. The entries in this table indicate the wide range of conditions under which thermal bremsstrahlung may be important.

To compute the contribution from all the various ions in the plasma it is necessary to perform the summation

$$S = \sum N_E N_Z Z^2 \qquad (5\text{-}60)$$

This requires a knowledge of the ionization state of the gas and of the abundances of the elements. For the cosmic abundances of the elements (Aller, 1961) about 90% of the ions are hydrogen, about 9% helium and about 1% heavier elements. The main contribution to the sum comes from hydrogen and helium. For a fully ionized plasma (temperatures above 10^5 °K, roughly) we have, taking into account the electric neutrality of the plasma ($N_e = \sum Z N_Z$)

$$S = 1.4 \, N_e^2 \qquad \text{(cosmic abundances)} \qquad (5\text{-}61)$$

Table 5.3. Characteristics of Some Probable Low Frequency Bremsstrahlung Sources

Source	Frequency of Observed Bremsstrahlung Spectrum	Electron Density (cm^{-3})	Temperature (^{o}K)
Solar Flare	Radio-microwave X-ray	10^{10}	10^{7}
HII Region	Radio	10-100	10^{5}
Orion	Radio	700	10^{4}
Sco X-1	Optical-X-ray	10^{16}	10^{8}
Coma Cluster	X-ray	10^{-3}	10^{8}

For a transparent medium the intensity of the bremsstrahlung radiation is proportional to the emission integral:

$$\langle N_e^2 \; V\rangle = \int N_e^2 \; dV \tag{5-62}$$

where V is the volume of the source. For extended sources the corresponding emission along the line of sight is what is measured. Then the emission is proportional to the emission measure:

$$\langle N_e^2 \; R\rangle = \int N_e^2 \; dR \tag{5-63}$$

To compute the total bremsstrahlung emission, Equation (5-58) must be integrated over all frequencies.

$$j(T) = (32/3^{3/2})\alpha_f \; r_o^2 c(c/\bar{v}_{th})KT \; g_{ff}(T)N_e N_Z Z^2$$

$$= 1.4\times10^{-27} \; T^{\frac{1}{2}} \; S \; g_{ff}(T) \quad \text{erg/cm}^3 \text{ sec} \tag{5-64}$$

The results of Karzas and Latter show that $g_{ff} = 1.2 \pm 10\%$ for all temperatures greater than 10^6 °K. Using this value of g_{ff} in (5-64) and setting $S = 1.4 \; N_e^2$, we find

$$j(T) = 2.4\times10^{-27} \; T^{\frac{1}{2}} \; N_e^2 \qquad \text{erg/cm}^3 \text{ sec.} \tag{5-64$'$}$$

For a plasma having a composition similar to that of the cosmic abundances bremsstrahlung is the principle radiative energy loss mechanism at temperatures above 10 million degrees. From Equation (5-64) the cooling time of such a plasma is therefore

$$t_{ff} = 3\ N_e KT/j(T) = 1.8\times10^{11}\ T^{1/2}/N_e \quad \text{sec} \tag{5-65}$$

5.7. Free-Free Absorption

When an electron collides with an ion it is also possible for the electron to absorb rather than emit a photon and make a transition to a higher rather than lower energy state. The absorption coefficient for this process, which is called inverse bremsstrahlung or free-free absorption, can be readily calculated from Equation (5-58) and (2-151) for a Maxwellian distribution of electron velocities. Averaged over solid angle, it is

$$\mu(\nu) = (c^2/2h\nu^3)(e^{h\nu/KT} - 1)(j(\nu,T)/4\pi)$$

$$= 3.7\times10^{8}\ \nu^{-3}(1- e^{-h\nu/KT})T^{-1/2}$$

$$\cdot\ N_e N_Z\ Z^2\ g_{ff}(\nu,T)\ \text{cm}^{-1} \tag{5-66}$$

Because of the high dependence on the frequency, free-free absorption is important for the absorption of X-rays only when the matter is extremely dense, as in the interiors of stars. For low frequencies it can be important for fairly low densities and probably determines the shape of the low frequency spectrum in many thermal radio sources. For low frequencies ($h\nu << KT$)

$$\mu(\nu) = 0.018\times Z^2 N_e N_Z\ Z^2\ g_{ff}(\nu,T)/\nu^2\,T^{3/2}\ \text{cm}^{-1} \tag{5-66$'$}$$

5.8. Many-Body Effects, Bremsstrahlung From Electron-Atom and Electron-Electron Collisions

Many-Body Effects. So far we have considered only the bremsstrahlung produced in electron collisions with an isolated ion, and have neglected the influence of the interactions of the surrounding electrons and ions. For small impact parameters this neglect does not introduce an appreciable error, but when the impact parameter is large, the Coulomb potential of the target ion will be strongly modified. As discussed by Gould (1970) and Scheuer (1960), the maximum impact parameter for which the pure binary collision formula can be applied is not the interparticle spacing but the generally much larger Debye length:

$$b_{max} = \lambda_D = (kT/4\pi\ N_e\ e^2)^{\frac{1}{2}} \qquad (5\text{-}67)$$

This follows because the ions can be assumed to move independently of one another for distances less than the critical correlation length of a plasma, namely the Debye length.

The requirement for the validity of the results derived in the preceding sections is that the maximum impact parameter used there, $b_{max} = v/\omega$, be much less than the Debye length. However, since v/λ_D is the plasma frequency ω_p this requirement becomes simply

$$\nu >> \nu_p \qquad (5\text{-}68)$$

Electron-Atom Collisions. An electron undergoes a deflection or acceleration when it collides with an atom as well as an ion, and the dipole moment has a non-vanishing second time derivative, so bremsstrahlung will be produced. The ratio of the emissivity due to electron-atom bremsstrahlung to that due to electron ion bremsstrahlung is approximately

$$j_{e-a}/j_{e-i} \simeq (R_a/b_o)^2 (N_a/N_i)$$
$$\simeq (T/10^5 Z^2)^2 \ (N_a/N_i) \qquad (5\text{-}69)$$

where R_a is the radius of the atom, $b_o = Ze^2/m_e v^2$, and N_a and N_i are the electron and ion number densities. Thus, for example, in a hydrogen gas that is 50% ionized and has a temperature of 10^4 °K, the ratio of the electron-atom to the electron-ion emissivity is on the order of 1% (Gould (1969), Bekefi (1966)).

Electron-Electron Bremsstrahlung. As discussed at the beginning of this chapter, electron-electron collisions produce no bremsstrahlung in the dipole approximation. However, there is a quadrupole contribution, which is on the order $(v/c)^2$ times the dipole term (see Sections 1.9 and 2.8).

Calculations using the Born-Elwert cross section (5-53) show that the ratio of the quadrupole electron-electron emission to the dipole electron-ion emission from a Maxwellian plasma is (Gould, 1975)

$$j_{e-e}/j_{e-i} \simeq (N_e/Z^2 N_Z)(KT/(m_e c^2 + 0.79KT)) \qquad (5\text{-}70)$$

5.9. Relativistic Bremsstrahlung

In the relativistic case the energy of the

electron must be written as

$$W = (p^2c^2 + m_e^2c^4)^{\frac{1}{2}} \tag{5-71}$$

Dirac wave functions must be used instead of Schrodinger wave functions, and in the matrix element (5-37) we must replace the non-relativistic operator for the momentum by the Dirac operator $\vec{\alpha}$ (see Equation (2-142)). When the Born approximation applies, namely, when the initial and final kinetic energies of the electron are both much greater than their potential energy in the Coulomb field, the cross section can be expressed in closed form (see Heitler, (1954)). In the case where the final momentum of the electron p_f is much greater than m_ec, the cross section is (Heitler (1954), Bethe and Salpeter (1957), Feynman (1962))

$$\nu\sigma(\nu) = 4Z^2\alpha_f r_o^2 (1+(p_f/p_i)^2 - (2p_f/3p_i))$$

$$\cdot (\log(2p_f p_i/h\nu m_e) - \tfrac{1}{2}) \tag{5-72}$$

When p_f is much less than m_ec

$$\nu\sigma(\nu) = 2Z^2 \alpha_f r_o^2 (p_f/m_ec) \tag{5-73}$$

This cross section tends to zero at the hard photon limit where $p_f = 0$. If exact wave functions were used it would probably tend to a finite limit as in the non-relativistic case (see Table 5.1).

For highly relativistic electrons the screening of the nucleus by atomic electrons becomes important. In the absence of screening the maximum impact parameter is

$$b_{max} \simeq \gamma^2 v/\omega \qquad (5\text{-}74)$$

where the additional factor γ^2 comes from the relativistic contraction of the electric fields and the Doppler effect (see Jackson (1962)). When b_{max} is greater than the atomic radius, atomic electrons will screen the nucleus and greatly reduce the cross section. If we use the Thomas-Fermi model for the atom, then the atomic radius is on the order of

$$R_a \sim a_o Z^{-1/3} \qquad (5\text{-}75)$$

(a_o is the Bohr radius) and screening is important when

$$\gamma^2 >> 137\ Z^{-1/3}\ (h\nu/m_e c^2) \qquad (5\text{-}76)$$

For small Z it is more appropriate to use the simple Bohr model for the atom wherein $R_a = a_o/Z$, so the factor $Z^{-1/3}$ replaced by a factor Z^{-1}.

In the limit of complete screening the factor

$$\log(2p_i p_f/h\nu m_e) - \frac{1}{2} \rightarrow \log(183Z^{-1/3}) + (p_i/9p_f) \qquad (5\text{-}77)$$

In an ionized gas screening becomes important when b_{max} becomes greater than the Debye length (5-67). Then

$$\log(2p_f p_i/h\nu m_e) \rightarrow \log(2\lambda_D\, m_e c/\hbar) \qquad (5\text{-}78)$$

The radiative energy loss rate in the extreme relativistic limit is given by

$$dW/dt = 4Z^2 \alpha_f r_o^2 c N_Z W_i \begin{cases} \log(2p_i/m_e c) - (1/3); \\ \quad \gamma \ll 137Z^{-1/3} \\ \log(183Z^{-1/3}) + (2/9); \\ \quad \gamma \gg 137Z^{-1/3} \end{cases} \qquad (5\text{-}79)$$

The ratio of the bremsstrahlung to ionization losses is (cf. Section 5.1)

$$(dW/dt)_B/(dW/dt)_I \simeq 10^{-3}(Z\ W_i/m_e c^2) \qquad (5\text{-}80)$$

so that for energies greater than about 100 MeV, bremsstrahlung losses dominate.

Finally, in the relativistic regime, the bremsstrahlung in the field of the atomic electrons must be taken into account. For electrons whose energies are much greater than $m_e c^2$, but not so large that screening is important, the cross section for electron-electron bremsstrahlung is the same as for electron-ion bremsstrahlung, except that the factor Z^2 in Equation (5-72) must be replaced by Z. This follows because the Coulomb potential is e^2/r rather than Ze^2/r, so the effective acceleration and hence the emission is down by a factor Z^2. This must be multiplied by the number of electrons per ion, Z. Calculations of electron-electron bremsstrahlung in the complete screening limit show that the screening reduces the radiation in the field of an electron less than that in the field of the nucleus. The effect is that the factor $183Z^{-1/3}$ in the logarithmic term is replaced by $1440Z^{-2/3}$ (Wheeler and Lamb (1939)).

General References And Suggested Reading

5.1. Collisional Or Ionization Losses

Jackson (1962)

Rossi (1952)

Bethe and Ashkin (1953)

Gould (1972 a,b)

5.2. Bremsstrahlung-Classical Calculation

Landau and Lifshitz (1962)

Bekefi (1966)

Gould (1970)

5.3. Classical Bremsstrahlung--Order of Magnitude Estimates

Jackson (1962)

Novikov and Thorne (1973)

5.4. Bremsstrahlung--Non-Relativistic Born Approximation

Heitler (1954)

Frank-Kamenetskii (1962)

Bethe and Salpeter (1957)

5.5. Bremsstrahlung--Exact Non-Relativistic Results

Frank-Kamenetskii (1962)

Sommerfeld (1939)

Elwert (1939, 1948)

Karzas and Latter (1961)

5.6. Bremsstrahlung From A Maxwellian Plasma

Karzas and Latter (1961)

Frank-Kamenetskii (1962)

5.7. Free-Free Absorption

Frank-Kamenetskii (1962)

Bekefi (1966)

5.8. Many-Body Effects, Electron-Atom and Electron-Electron Collisions

Gould (1970)

Bekefi (1966)

Gould (1975)

5.9. Relativistic Bremsstrahlung

Heitler (1954)

Bethe and Ashkin (1953)

Blumenthal and Gould (1970)

Feynman (1962)

Jackson (1962)

Problems

5.1. Compute the energy loss as a function of energy due to (a) ionization, (b) bremsstrahlung, (c) Compton scattering off the 3° black body radiation field, and (d) synchrotron losses for a relativistic electron in (i) the interstellar medium with a density of 1 hydrogen atom or ion per cubic centimeter and a magnetic field of 10^{-6} gauss, and (ii) the intergalactic medium with a density of 10^{-5} hydrogen ions per cubic centimeter and a magnetic field of 10^{-8} gauss.

5.2. The planetary nebula IC418 has a radius of 1.5×10^{17} cm, an electron density of 10^4 cm^{-3} and a temperature of 10^4 °K. Compute the bremsstrahlung intensity at 10^{10} Hz. At what frequency does free-free absorption become important? (See Aller and Liller (1968)).

5.3. Consider a solar flare in which 10^{25} ergs are emitted in the form of non-thermal bremsstrahlung X-rays with energies greater than 20 keV. Assuming that all the energy of the non-thermal electrons is deposited in the chromosphere by means of collisional losses, and subsequently re-emitted in the far UV range, estimate the far UV energy associated with the non-thermal X-ray burst. (For a detailed computation, see Petrosian (1973) and references.)

5.4. The Coma cluster X-ray source has a 2-10 keV X-ray luminosity of 5×10^{44} erg/sec, a temperature of 10^8 °K, and a radius of

2×10^{24} cm. Assuming a uniform distribution of matter, compute the electron density, the mass of the X-ray emitting plasma, its thermal energy content and cooling time.

5.5. The X-ray star ScoX-1 emits about 10^{37} ergs/sec in X-rays above a kilovolt. Assuming that this emission is due to a uniform density and temperature plasma with a temperature of 10^8, and a density of 10^{16} cm^{-3}, compute (i) the radius of the plasma, (ii) its optical luminosity, and (iii) the frequency at which free-free absorption becomes important. (Remember to take into account the effects of electron scattering on the absorption coefficient. See Equation (4-62.)

5.6. Compute the shape of the bremsstrahlung spectrum from a non-isothermal plasma in which (i) $T \propto R^{-2}$, $N \propto R^{-3/2}$, (ii) NT = constant.

Chapter 6

RADIATIVE RECOMBINATION

6.1. Radiative Recombination Rate Coefficient

In bremsstrahlung a free electron of energy W_i encounters an ion of charge Ze and emits a photon of energy $h\nu$ while making a transition to another free state W_f. From energy conservation the energy of the photon is $h\nu = W_i - W_f$. If instead the electron makes a transition to a bound state of energy $-I_{Z,n}$ the electron is said to have recombined with the ion. In order to conserve energy, a photon of energy

$$h\nu = W_i + I_{Z-1,n} \tag{6-1}$$

must be emitted. Since W_i is a continuous positive variable, the photon energy can assume any value greater than $I_{Z-1,n}$. Therefore the recombination spectrum is continuous, with edges or discontinuities at $W = I_{Z-1,n}$. To calculate the spectrum it is necessary to consider the recombination to the individual levels $I_{Z-1,n}$.

In the limit where W_f and $I_{Z-1,n} \to 0$ the cross section for bremsstrahlung and recombination should be equal. This fact can be used to obtain an expression for the recombination cross section to hydrogenic ions which turns out to be accurate even for captures to the ground state (Gould and Thakur (1970)). Setting

$$\Delta\sigma_B = \sigma_B(\nu)\Delta\nu = \Delta\sigma_R \tag{6-2}$$

with

$$h\Delta\nu = \Delta I_{Z-1,n} = 2Z^2 \; I_H(\Delta n/n^3) \tag{6-3}$$

yields

$$(\Delta\sigma_R/\Delta n) = \sigma_R(n) = (32\pi/3\sqrt{3}) \; Z^2 \alpha^3 k_e^{-2}$$
$$\circ \; (Z^2 I_H/h\nu)(g_R(n)/n^3) \tag{6-4}$$

where $g_R(n)$ is the recombination Gaunt factor which is = 1 within 10% (Karzas and Latter (1961); Glasco and Zirin (1964)), and I_H is the ionization potential of hydrogen.

The total recombination coefficient is

$$\alpha_r = \sum_n \langle \sigma(n) v_i \rangle \tag{6-5}$$

where the average is over a Maxwellian distribution of incident electron velocities v_i. For large n, the cross section is proportional to n^{-3}. For low electron velocities, or energies $W_i << Z^2 I_H/n^2$, the photon energies $h\nu \simeq Z^2 I_H/n^2$, and the cross section is proportional to n^{-1}. Thus the dependence of the cross section on the principal quantum number n is roughly as follows:

$$\sigma(n) \propto \begin{cases} n^{-1} \quad ; & n \leq n_{max} \\ n^{-3} \quad ; & n \geq n_{max} \end{cases} \tag{6-6}$$

with

$$n_{max} \simeq (Z^2 I_H/W_i)^{\frac{1}{2}} \simeq (Z^2 I_H/KT)^{\frac{1}{2}} \tag{6-7}$$

Therefore as long as $W_i \simeq KT << Z^2 I_H$

$$\alpha_r \propto \langle v_i/W_i \rangle \sum_n n^{-1} \propto T^{-\frac{1}{2}} \ln(Z^2 I_H/KT) \qquad (6\text{-}8)$$

The expression for α_r obtained by summing over all n and averaging over a Maxwellian distribution of v_i is (Spitzer, 1948, Burbidge, Gould and Pottasch, 1963).

$$\begin{aligned}\alpha_r &= 2A(2KT/\pi m)^{\frac{1}{2}}\, y\, \varphi(y)\overline{g} \\ &= 2\times10^{-11}\, Z^2 T^{-\frac{1}{2}}\, \varphi(y) \quad \text{cm}^3\,\text{sec}^{-1}\end{aligned} \qquad (6\text{-}9)$$

where

$$A = (32\pi/3^{3/2})\alpha_f^3\, a_o^2 = 2\times10^{-22}\ \text{cm}^2 \qquad (6\text{-}10)$$

and

$$y = Z^2 I_H/KT \qquad (6\text{-}11)$$

and $\overline{g}$ is an average Gaunt factor = 0.9 to within 10% (Gould and Thakur, 1970). For low temperatures φ is given by (Burbidge _et al_., 1963)

$$\varphi(y) = 1/2(1.735 + \ell n y + (6y)^{-1})\,;\ y \gtrsim 1 \qquad (6\text{-}12)$$

For small y the leading term is 1/2 ln y, so the temperature dependence of α_r is as given in Equation (6-8). Although Equation (6-9) was derived in the limit of large y it is accurate to better than 1% even for y as small as 0.5. For small y (high T) one can derive an expression for φ which is a power series in y. The first term in this series is

(Tucker and Gould, 1966)

$$\varphi(y) = y(-1.202\,\ell n y - 0.298)\ ;\ y \ll 1 \tag{6-13}$$

so at high temperatures

$$\alpha_r \propto Z^4 T^{-3/2}\ \ell n(KT/Z^2 H)\ ;\ KT \gg Z^2 I_H \tag{6-14}$$

The recombination rate decreases at high temperatures partly because of the reduced relative contribution of radiative captures to excited states.

In some astrophysical situations the photon absorption mean free path is small compared to the dimensions of the system under consideration, so radiative captures to the ground state, which produce a photon of energy greater than the ionization potential, just ionize another atom. In this case, which is called Case B in the literature, the recombinations to the ground state do not result in a net change in the ionization balance of the plasma. The appropriate recombination coefficient is one in which the summation is over excited states only. It is the same as (6-9) but with φ replaced by $\varphi' = \varphi - \varphi_1$ where φ_1 represents the contribution from capture to the ground state:

$$\varphi_1 = ye^y K(y) \rightarrow \begin{cases} 1 - y^{-1} - 2y^{-2}\ , & \text{large } y \\ y(-\ell n\, y - 0.577), & \text{small } y \end{cases} \tag{6-15}$$

where K(y) is the exponential integral (Gould and Thakur, 1970).

6.2. Radiative Recombination Rate Coefficient For Complex Ions

For non-hydrogenic ions the contribution to the total rate from captures to excited states may be computed with sufficient accuracy from the hydrogenic formulas since the excited states are approximately hydrogenic. For example, the recombination rate to states with n greater than 3 for Lithium-like ions may be computed from Equation (6-9) with $\varphi(y)$ replaced by $\varphi'' = \varphi - \varphi_2$ where φ_2 includes the n = 1 and n = 2 contributions. For recombinations to the unfilled valence shells, a fair approximation is obtained by replacing the principal quantum number n by the effective quantum number n^*, where

$$n^* = z(I_H/I_Z)^{\frac{1}{2}} \qquad (6\text{-}16)$$

where z is the effective nuclear charge, which assumes perfect screening by the inner shell electrons. The presence of electrons in the valence shell is accounted for by multiplying the rate by the effective fraction of the shell that is empty ($\overline{s}/2n^2$). Thus, for example, in the recombination of a Lithium-like ion to form a Beryllium-like ion, there would be only one electron out of a possible eight in the 2n shell, so the rate would be multiplied by a factor (8-1)/8 = 7/8. Cox and Tucker (1969) have given a more detailed treatment along these lines.

A more accurate treatment results from using the fact that the free-bound recombination process is just the inverse of the bound-free photoionization process. Therefore, recombination coefficients can be obtained from photoionization coefficients when they are known, by using the detailed balance procedure similar to that described in Section 2.2. Gould and Thakur (1970) estimated the ground state recombination coefficient for singly ionized helium in this manner. They found $\varphi_1 = 0.82$ to be compared with a value of $\varphi_1 = 0.85$ which follows from the rough method described above.

Tarter (1971, 1973) has used the detailed balance method to obtain a 3-term parametric expression which agrees with detailed computations to better than 3%.

6.3. Dielectronic Recombination

For complex ions at fairly high temperatures and low densities, dielectronic recombination will be the most important process by which an electron can recombine with an ion (Burgess, 1964). In this process an electron incident on an ion of net charge z collisionally excites the ion to an upper level and is at the same time captured to some other excited state. Thus the process is analogous to radiative recombination which can be viewed as bremsstrahlung collisions in which the electron is captured. In the dielectronic process the recombined ion is doubly excited with a total energy lying in the continuum of the ion z-1. This system is unstable to autoionization (Auger effect) and usually takes this path. However, occasionally the ion will be stabilized by a radiative transition to a lower level and the recombination will be effected. Because of the necessity of supplying energy to excite the ion z, only fairly energetic incident electrons can recombine this way. When averaged over a Maxwellial distribution this introduces a factor exp(-ΔW/KT) where ΔW is the excitation energy. The total rate can be written roughly as

$$\alpha_d \simeq C\, T^{-3/2}\, e^{-\Delta W/KT}\, N_e N_z \qquad (6\text{-}18)$$

where C is in the range 10^{-3} to 10^{-1}, being generally larger for helium-like ions with z on the order of 5 or greater. For all ions with small z (on the order of unity) C is small. (Burgess, 1965, Shore, 1969, Beigman, Vainshtein, and Vinogradov, 1970) Comparison of Equation (6-18) with Equation (6-9) for the radiative recombination rate shows that dielectronic recombination will dominate over radiative recombination at temperatures $KT \lesssim 0.3\ \Delta W$.

For high particle or photon densities the dielectronic recombination rate will be reduced by collisional and photoionization out of the highly excited state. Burgess and Summers (1969) have shown that electron densities in excess of 10^{12} cm^{-3} will cause an order of magnitude decrease in the rate, whereas radiation fields as intense as the photospheric radiation field (T_r = 5,600, see Equation (2-27)) have a negligible effect.

6.4. Continuum Radiation Spectrum

For a Maxwellian distribution of electron velocities the continuous recombination radiation spectrum due to capture into level n of the ion z-1 is

$$j(\nu) = N_Z h\nu\ \sigma_R(n) N(v) v (dv/d\nu)$$

$$= 1.8\times10^{-32}\ T^{-3/2}\ n^{-3}\ N_e N_Z Z^4$$

$$\cdot \exp[(I_{Z-n,n} - h\nu)/KT] \quad \text{erg/cm}^3\text{ sec Hz}$$

(6-19)

For non-hydrogenic ions the ionization energy is $I_{Z,z,n}$ where Z is the nuclear charge and z is the net ionic charge. The spectrum due to recombination to non-hydrogenic ions may be approximated by setting $Z^2 I_H/n^2 = I_{Z,z-1,n}$ and multiplying Equation (6-19) by a factor $(\overline{s}/2n^2)$ which represents the incompleted fraction of shell n. To obtain the spectrum of a plasma consisting of a mixture of ions Z and states of ionization z, Equation (6-19) must be summed over all ions and all levels n for which $I_{Z,z-1,n} > h\nu$. The result is

$$j(\nu) = 1.8\times10^{-32}\ N_e N_H T^{-3/2}\ e^{-h\nu/KT}\ X$$

$$\text{erg/cm}^3\text{ sec Hz} \qquad (6\text{-}20)$$

where

$$X = \sum_{Z,z,n} (N_{Z,z}/N_Z)(N_Z/N_H)(\overline{s}/2n^2)\, n (I_{Z,z-1,n}/I_H)^2$$

$$\cdot \exp(I_{Z,z-1,n}/KT) \qquad (6\text{-}21)$$

In Equation (6-21) $N_{Z,z}$ is the density of ions with nuclear charge Z and net ionic charge z, and the sum is over all ions and all levels n for which $I_{Z,z-1,n} > h\nu$. At any given temperature only a few terms make an appreciable contribution to the sum. Comparison of Equations (6-20) and (5-58) shows that the ratio of recombination to bremsstrahlung radiation at a given frequency is

$$j_R(\nu)/j_B(\nu) \approx 10^{-1} X/T_6 \qquad (6\text{-}22)$$

where T_6 is the temperature in millions of degrees. For a given temperature T_6 recombination dominates bremsstrahlung at wavelengths below about $30/T_6$ Å. For wavelengths larger than this value, bremsstrahlung dominates. For temperatures above 10^7 °K, recombination is unimportant at all wavelengths, except for the edges at $I_{Z,z-1,n} = h\nu$ (see Equation (6-1)), which may be detectable features in the spectrum (Culhane, 1969; Tucker and Koren, 1971).

6.5. Balmer Lines

In addition to a continuous spectrum, radiative recombination can result in the production of emission lines as the electron cascades to lower levels following capture in an excited state. The cascades to the n = 2 level result in the Balmer series of lines which is of considerable importance in astrophysics. The Hα and Hβ lines are an important feature of the optical spectrum of any low density gas having a temperature in the range 10^3-10^5 °K.

By solving the capture cascade equations Pengelly (1964) has calculated the emissivity of the Balmer lines. A convenient analytic expression for the emissivity of a Balmer Hn line as a function of temperature is

$$j(H{:}n,2) = 10^{-23}\ N_e N_p A_{n2}\ T^{-\frac{1}{2}} \log(I_H/KT)$$

$$\text{erg/cm}^3\ \text{sec} \qquad (6\text{-}23)$$

where for Case B recombination (ground state recombinations cancelled out by photionization)

$$A_{32}(H\alpha) = 1.29\ ;\quad A_{42}(H\beta) = 0.449$$

$$A_{52}(H\gamma) = 0.209\ ;\quad A_{62}(H\delta) = 0.115 \qquad (6\text{-}24)$$

(Burbidge *et al.*, 1963, Delmer, Gould and Ramsay, 1967.)

For the strong recombination lines of He(λ4471) and He^+(λ4686), Delmer, *et al.*, (1967) find

$$j(HeI,\lambda 4471) = 5.18\times 10^{-26}\ N_e N_{He^+}\ (T=10^4\ {}^oK)$$

$$j(HeII,\lambda 4686) = 1.58\times 10^{-24}\ N_e N_{He^{++}}$$

$$\text{erg/cm}^3\ \text{sec} \qquad (6\text{-}25)$$

The temperature dependence should be about the same as for the hydrogen Balmer lines.

The above formulas are for temperatures less than about thirty thousand degrees. For higher temperatures the dependence of the Balmer lines should be roughly as $T^{-3/2}$ $\log(KT/I_H)$ (Gould, 1971; cf. Equation (6-14)).

6.6. Radiofrequency Recombination Lines

The Balmer lines arise from the transitions into the n = 2 level following recombination to an excited state. Other series of lines will be produced by transitions into other levels. Apart from the Balmer and Lyman (n = 1) series, the transitions which have been the most important for astrophysical applications are those between levels of very large principal quantum number n. These lines, which result in radio frequency line emission, were first predicted by Kardashev (1959). They provide important information on the behavior of atoms in highly excited states, and the physical conditions in the interstellar medium.

The frequency of the lines are given by

$$\nu = cRy\ Z^2 (n_1^{-2} - n_2^{-2}) \qquad (6\text{-}26)$$

where Ry is the Rydberg constant ($cRy \equiv I_H$) taken for an excited atom of net charge z as seen by the electron in the highly excited state. If we take into account the fact that the reduced mass of the electron is slightly less than the rest mass, the Rydberg constant is found to be

$$Ry = (2\pi^2 m_e e^4 / ch^3)/(1 + m_e/Am_p) \qquad (6\text{-}27)$$

where A is the atomic weight and m_p is the proton mass. The variation of the Rydberg constant with mass of the atom is illustrated in Table 6.2.

When $n_2 = n_1+1$, the line is denoted as an $n\alpha$ transition: $n_2 = n_1+2$, transitions are $n\beta$ transitions, etc. For most purposes it is accurate enough to approximate equation (6-26)

Table 6.2. Variation of the Rydberg Constant With Atomic Number

Element	Atomic Weight	Ry(A)
H^1	1.007	109678 cm^{-1}
H^2	2.00	109707
He	4.00	109722
C	12.00	109733
O	16.00	109733

by

$$\nu \simeq 2\ c\ Ry\ Z^2\ (n_2 - n_1)/n_1^3 \qquad (6\text{-}28)$$

Radio recombination lines have been observed from hydrogen, helium and carbon in the frequency range between about 400 and 10,000 MHz, corresponding to values of n in the range 50 to 250.

The emissivity at the center of the line is made up of contributions from both line and continuum. If we assume that the emitting volume is isothermal and of uniform density, the intensity at the center of the line is given by (see Equation (2-148))

$$I = (j(\nu)/\mu(\nu))(1 - e^{-\tau(\nu)}) \qquad (6\text{-}29)$$

This intensity consists of two parts,

continuum and line:

$$I = I_L + I_C \tag{6-30}$$

Thus, the ratio of the line intensity to the continuum intensity is given by

$$I_L/I_C = (I/I_C) - 1 = \frac{(j/\mu)}{(j_c/\mu_c)} \frac{(1-e^{-\tau(\nu)})}{(1-e^{-\tau_c(\nu)})} - 1 \tag{6-31}$$

The lifetime of the excited states is usually long compared to the collision time, so it is often a good approximation to assume that the population of the levels are given by the conditions of local thermodynamic equilibrium (LTE). In this case the ratio $j(\nu)/\mu(\nu) = B(\nu)$ where $B(\nu)$ is the Planck function (see Equation (2-151)), and

$$I_L/I_C = (1- e^{-\tau(\nu)})/(1- e^{-\tau_c(\nu)}) - 1 \tag{6-32}$$

For an optically thick gas $\tau(\nu)$ and $\tau_C(\nu) >> 1$, so $I_L/I_C = 0$, i.e., the line vanishes. For the optically thin case,

$$I_L/I_C = (\tau(\nu)/\tau_C(\nu)) - 1 = \tau_L(\nu)/\tau_C(\nu) \tag{6-33}$$

The optical depth for the continuum is taken from Equation (5-66′) for free-free absorption. The optical depth at the line center is (see Problem 8.5)

$$\tau_L \simeq 1\times10^7 \; Z^2 \,(f(n_1, n_2)/n)\, T^{-5/2} \; N_e N_p R_{pc}/\Delta\nu_L \tag{6-34}$$

where R_{pc} = size of nebula in pc and $\Delta\nu_L$ is the full line width at half power. Since Doppler broadening is the cause of the line width (see Equation (8-65))

$$\Delta\nu_L = \Delta\nu_o(\log 2)^{\frac{1}{2}} = 3.4\times10^{-7}\ T^{\frac{1}{2}}\ A^{-\frac{1}{2}}\ \nu \qquad (6\text{-}35)$$

For large n, the oscillator strength $f(n_1, n_2) \simeq 0.2$ n so for a hydrogen $n\alpha$ transition in an optically thin gas (Dupree and Goldberg, 1970)

$$I_L/I_C \simeq 1.0\times10^{-4}\ \nu\ T^{-3/2}\ g_{ff}^{-1}\ (N_eN_p/S) \qquad (6\text{-}36)$$

where S is the bremsstrahlung sum given by Equation (5-60). For most cases of interest $S = 1.2\ N_eN_p$ to within about 10%. Detailed calculations (see Dupree and Goldberg, 1970, and references cited therein) show that departures from LTE are small for n greater than about 150, but are quite important for n as low as 50. At low densities, the level populations can become inverted and stimulated emission can become important. This is especially true for a complex atom such as carbon where dielectronic recombination is an important process for populating the excited states.

General References And Suggested Reading

6.1. Radiative Recombination Rate Coefficient

Gould and Thakur (1970)

Bethe and Salpeter (1957)

Burbidge, Gould and Pottasch (1963)

Spitzer (1948)

Seaton (1959)

6.2. Rate Coefficient For Complex Ions

Tarter (1971, 1973)

Cox and Tucker (1969)

Elwert (1954)

Aldrovandi and Pequignot (1973)

6.3. Dielectronic Recombination

Burgess (1965)

Shore (1969)

Burgess and Summers (1969)

Beigman, Vainshtein and Vinogradov (1970)

6.4. Continuum Radiation Spectrum

Elwert (1954)

Culhane (1969)

Tucker and Koren (1971)

6.5. Balmer Lines

Delmer, Gould and Ramsay (1967)

Pengelley (1964)

Aller and Liller (1968)

6.6. Radiofrequency Recombination Lines

Dupree and Goldberg (1970)

Pacholczyk (1970)

Problems

6.1. Consider a source of ionizing radiation of luminosity L which is embedded in a uniform cloud of hydrogen of density N. The source will carve out an ionized region of radius R_i (Stromgren sphere). Beyond R_i the gas will be predominantly neutral, since the recombination rate becomes equal to the ionization rate. Assuming that the cloud is optically thick to the ionizing radiation, R_i can be estimated by equating the number of recombinations per second inside R_i to the rate of emission of ionizing photons. Show that $R_i = (3L/4\pi\ I\alpha\ N^2)^{1/3}$. Should Case A or Case B recombination coefficients be used? If the ionizing source is turned on suddenly, estimate the time taken for the formation of the Stromgren sphere (Mathews and O'Dell, 1969).

6.2. Compute the magnitude of the jump in the continuous spectrum produced by recombinations into the n = 2 level of hydrogen (Balmer discontinuity) for a nebula having a temperature of 10^4 °K, a density of 10^4 cm^{-3} and a radius of 3×10^{17} cm (see Aller and Liller, 1968).

6.3. Derive expressions for the relative abundances of HII, HeII and HeIII in terms of the Balmer line intensities.

6.4. Observations of the Coma Cluster show that the upper limit on the H emission from the diffuse gas is 1.5×10^{-16} ergs/cm^2 sec (arcsec)^2. Use this limit, together with a diameter of 8 Mpc to set a limit on the amount of matter at 3×10^4 ^oK (Goldsmith and Silk, 1972).

6.5. In the Orion Nebula the ratio $(I_L/I_C)\Delta\nu_L = 3\times10^4$ for the 100α transition in hydrogen. Assuming that the optically thin LTE calculation is valid, derive the temperature of the nebula (Dupree and Goldberg, 1970, Hjellming and Davies, 1970).

Chapter 7

THE PHOTOELECTRIC EFFECT

Consider an atom in some bound state with ionization potential I. If a photon of energy $h\nu$, such that

$$h\nu > I \tag{7-1}$$

is incident on the atom, an absorption process can occur in which one of the atom's electrons is ejected into a state of positive energy in the continuum. The energy of the ejected electron is given by

$$W = h\nu - I \tag{7-2}$$

Since W is continuous, absorption is possible for a continuous range of frequencies ν. This process is called the photoelectric effect.

The transition probability per unit time for this process is given by Equation (2-112) except now we must take into account the fact that there are infinitely many electron states with energy W. This has the consequence that we must multiply (2-112) by a "density of final states" term dn/dW (see Equations (2-135) and (2-139)).

Normalizing the final states to one particle per volume $V = X^3$, the transition probability per unit time for absorption becomes

$$dw = (4\pi^2 e^2/h^3)(v/c)(I(\omega)/\omega^2)$$

$$\cdot \; |\langle f|e^{i\vec{k}\cdot\vec{r}}\; \hat{\ell}\cdot\vec{\nabla}|i\rangle|^2 \; d\Omega \tag{7-3}$$

The differential cross section for absorption of a quantum of energy $h\nu$ with the ejection of an electron of velocity v into the solid angle $d\Omega$ is

$$\frac{d\sigma}{d\Omega} = \frac{dw/d\Omega}{I(\omega)/\hbar^2\omega}$$

$$= (\alpha_f\, v/\nu)\,|\langle f|e^{i\vec{k}\cdot\vec{r}}\,\hat{\ell}\cdot\vec{\nabla}|i\rangle|^2/(2\pi)^2 \qquad (7\text{-}4)$$

7.1. The Born Approximation--Hydrogen Atom

A useful approximation for computing the matrix element in (7-4) is to assume that the ejected electron can be treated as a free particle, so that its wave function is given by

$$u_f = e^{i\vec{k}_f\cdot\vec{r}}\,(V)^{-\frac{1}{2}} \qquad (7\text{-}5)$$

This will be a good approximation if the momentum of the ejected electron is much greater than the momentum of a typical bound state electron, i.e., $p_f \gg p_{1s}$ for hydrogen. This is the same as requiring

$$(p_f^2/2m) = (\hbar^2 k_f^2/2m) \gg e^2/2a_0$$

or

$$k_f^2\, a_0^2 \gg 1 \qquad (7\text{-}6)$$

In terms of the radiation wave number, this condition is

$$ka_o \gg \alpha_f$$

or (7-7)

$$h\nu \gg I_H$$

In the Born approximation the matrix element in (7-4) becomes

$$\langle f|e^{i\vec{k}\cdot\vec{r}}\ \hat{\ell}\cdot\vec{\nabla}|i\rangle = \int e^{-i\vec{q}\circ\vec{r}}\ (\partial u_i/\partial x)d\ xdydz \qquad (7\text{-}8)$$

where

$$\vec{q} = \vec{k}_f - \vec{k} \qquad (7\text{-}9)$$

and we have assumed the incident radiation to be polarized in the x-direction. Substituting in the ground state wave function for hydrogen (see Chapter 8) and integrating by parts, we find, for the differential cross section for photoionization from the 1S-state of hydrogen.

$$\frac{d\sigma}{d\Omega} = \frac{32e^2}{mc\omega}\ \frac{(k_f a_o)^3\ \cos^2\alpha}{(1 + q^2 a_o^2)^4} \qquad (7\text{-}10)$$

where

$$\cos\alpha = (\vec{k}_f\cdot\vec{x})/(k_f x) = \sin\theta\ \cos\varphi \qquad (7\text{-}11)$$

Equation (7-8) may be simplified by noting that

$$q^2 = k_f^2\ (1- 2(k/k_f)\ \cos\theta + (k/k_f)^2) \qquad (7\text{-}12)$$

where θ is the angle between the direction of the photon and the ejected electron.

In the non-relativistic Born approximation we have

$$k/k_f \simeq v/2c << 1 \tag{7-13}$$

Hence, to first order in v/c

$$q^2 = k_f^2 \left(1 - \frac{v}{c} \cos \theta\right) \tag{7-14}$$

and

$$q^2 a_o^2 \approx k_f^2 a_o^2 >> 1 \tag{7-15}$$

so

$$\frac{d\sigma}{d\Omega} = \frac{32e^2}{mc\omega(k_f a_o)^5} \frac{\sin^2\theta \cos^2\varphi}{(1- \frac{v}{c} \cos\theta)^4} \tag{7-16}$$

Integrating over angles, retaining terms of order v/c the total cross section is

$$\sigma(1S) = 2^8 (\pi/3)\alpha_f a_o^2 (I/h\nu)^{7/2} \tag{7-17}$$

For an ns-state

$$\sigma(ns) = (2^8\pi/3)(\alpha_f a_o^2/n^3)(I_H/h\nu)^{7/2} \tag{7-18}$$

For np states, the effect of the Coulomb field of the nucleus must be taken into account. Using first order perturbation theory (see Bethe and Salpeter, 1957) it is found that the cross section for photoionization from a np state of hydrogen is

$$\sigma(np) = \frac{2^7 \pi}{9} \frac{n^2-1}{n^5} \alpha_f a_o^2 (I_H/h\nu)^{9/2} \tag{7-19}$$

Note that the photoionization cross section decreases rapidly with increasing n and ℓ.

7.2. The Dipole Approximation

Another useful approximation is to set $e^{i\vec{k}\cdot\vec{r}}$ equal to unity in the matrix element (7-4). This is the dipole approximation, encountered in the discussion of emission. It is a valid approximation provided

$$ka_o \ll 1 \tag{7-20}$$

or

$$h\nu \ll (2/\alpha_f) I_H \tag{7-20$'$}$$

This approximation can be used to obtain the cross section near threshold. Also in the region where $\alpha_f \ll ka_o \ll 1$, both the dipole and the Born approximation are valid. The differential cross section then becomes (see Equation (2-123))

$$d\sigma/d\Omega = (4\pi^2 \alpha_f m_e^2 \nu v/h^2)|\langle f|\hat{\ell}\cdot\vec{r}|i\rangle|^2 \tag{7-21}$$

Often the matrix element in (7-21) is expressed in terms of a final state wave function normalized to unit energy interval:

$$\int U_{W\ell m}(r)\, U_{W'\ell'm'}(r) d^3x = \delta_{\ell\ell'} \delta_{mm'} \delta(W-W') \tag{7-22}$$

The matrix element in (7-23) may be expressed in terms of a continuum oscillator strength normalized to unit energy interval (cf. Equation (2-132)):

$$df(f,i)/dW = (h\nu/I_H)\,|\langle f^W|\hat{\ell}\cdot\vec{r}|i\rangle|^2/a_0^2 \qquad (7\text{-}24)$$

Thus (see Equation (2-139))

$$d\sigma/d\Omega = 4\pi^2\ \alpha_f a_0^2\ I_H df(f,i)/dW \qquad (7\text{-}23')$$

The quantity $I_H df(f,i)/dW$ should be of the order unity at threshold, or the threshold photoionization cross section is on the order of

$$\sigma_{th} \sim a_0^2 \qquad (7\text{-}25)$$

Here $|f^W\rangle$ refers to a final state normalized to unit energy interval.

Inserting the bound state wave function for the 1s-state and using <u>exact</u> wave functions for the final state the cross section for photoionization from the 1s-state of hydrogen is found to be (see Bethe and Salpeter, 1957)

$$\sigma(1S) = \frac{2^9\pi^2}{3}\alpha_f\ a_0^2\ (I_H/h\nu)^4\ G(n_f) \qquad (7\text{-}26)$$

where

$$G(n_f) = e^{-4n_f \operatorname{arctg} n_f} / 1 - e^{-2\pi n_f} \qquad (7\text{-}27)$$

and (Equation (5-52))

$$n_f = (I_H/W_f)^{\frac{1}{2}} = Ze^2/\hbar v \qquad (7\text{-}28)$$

For an hydrogenic ion, assuming hydrogenic wave functions, Equation (7-26) applies, provided we make the replacements

$$I_H \rightarrow Z^2 I_H \qquad a_o \rightarrow a_o/Z$$

$$W_f \rightarrow Z^2 I_H/n_f^2 \qquad n_f \rightarrow (Z^2 I_H/W_f)^{\frac{1}{2}} \qquad (7\text{-}29)$$

In the Born-approximation limit,

$$G(n_f) \simeq (1/2\pi n_f) = (1/2\pi)(h\nu/Z^2 I_H)^{\frac{1}{2}} \qquad (7\text{-}30)$$

so

$$\sigma(1S) = (2^8\pi/3)\alpha_f\, a_o^2\, Z^{-2}(I/h\nu)^{7/2}$$

$$= 5.6\times10^{-17}\, Z^{-2}\, (I/h\nu)^{7/2} \qquad (7\text{-}31)$$

Near threshold $h\nu \simeq I$ and

$$G(n_f) \approx e^{-4}(1+ 4/3n_f^2) \simeq e^{-4}(h\nu/I)^{4/3} \qquad (7\text{-}32)$$

so

$$\sigma(1S) = 31\, \alpha_f\, a_o^2\, Z^{-2}\, (I/h\nu)^{8/3}$$

$$= 6.4\times10^{-18}\, Z^{-2}\, (I/h\nu)^{8/3} \qquad (7\text{-}33)$$

Near threshold, the cross section depends

on Z as $Z^{10/3}$. The frequency dependence in the various ranges of $h\nu/I$ is

$$\sigma(1S) \propto \begin{cases} \nu^{-8/3} & h\nu \simeq I \\ \nu^{-3} & h\nu \gtrsim I \\ \nu^{-7/2} & h\nu >> I \end{cases} \qquad (7\text{-}34)$$

7.3. K-Shell Photoionization Cross Sections

For non-hydrogenic atoms we can get a rough estimate of the K-shell ionization cross section by simply multiplying Equation (7-26) by a factor 2 to account for the presence of 2 1S electrons. However, this provides only a crude (factor 2) estimate. Better agreement with the experimental data is obtained by replacing Z by Z-S where S is a constant that takes some account of the screening of the nuclear potential by the electrons. A value of S = 0.66 provides a good (within 20%) fit to the exact calculations and the experimental data for both the threshold energies and cross sections (see Brown, 1971, Brown and Gould, 1970, Daltabuit and Cox, 1972, and references cited therein). Thus

$$\sigma_{th}(K, \text{He-like}) \cong 1.3\times10^{-17}\ (Z-0.6)^{-2} \qquad (7\text{-}35)$$

Daltabuit and Cox (1972) have used the experimental data of Henke to obtain an interpolation formula which gives the K-shell photoionization cross section for any ion to within 30%. They find for the threshold energy, $(h\nu)_{th}$

$$h\nu_{th} = Z^2_{eff}\ I_H \qquad (7\text{-}36)$$

where

$$Z_{eff} = Z\ N^{0.12-\ 0.22/\log_{10}Z} \qquad (7\text{-}37)$$

N is the number of electrons in the ion. The threshold cross section is

$$\sigma_{th}(K,N) = 2.2\times10^{-17}\ Z_{eff}^{-2} \qquad (7\text{-}38)$$

In addition they fit the data near threshold to a formula of the form

$$\sigma = \sigma_{th}(\nu_{th}/\nu)^{s} \qquad (7\text{-}39)$$

with

$$s = 2.0\quad (Z_{eff})^{0.14} \qquad (7\text{-}40)$$

The values of s obtained in this manner are all in the range 2.5-2.9, so they must be used with caution, i.e., only for photon energies not too far above threshold. For photon energies far above threshold s should approach the Born limit value s = 3.5.

The intensity of a monochromatic beam of ionizing radiation passing through matter of density N atoms/cm^3 decreases exponentially with distance R in the absence of any radiation by the matter (see Equation (2-148)):

$$I = I_o\ e^{-\mu R} \qquad (7\text{-}41)$$

where

$$\mu = N\sigma \qquad cm^{-1} \qquad (7\text{-}42)$$

is the absorption coefficient. When dealing

with high energy radiation the results are often expressed in terms of the mass absorption coefficient

$$\mu_m = \mu/\rho = \sigma/Am_p \qquad \text{cm}^2/\text{gm} \qquad (7\text{-}43)$$

7.4. K-Fluorescence Yield

When an electron is ejected from the K-shell of an atom or ion, the resulting ion is left in a highly excited state which is in the continuum. The atom may stabilize either by the emission of a photon or by the emission of an electron in the auto-ionization or Auger process. The probability that an electron will emit a K-line photon is roughly given by (Wentzel, 1927, McGuire, 1969)

$$p(K) = Z^4/(Z^4 + 33^4) \qquad (7\text{-}44)$$

7.5. X-ray Absorption Of The Interstellar Medium

The photoelectric effect has important consequences for X-ray astronomy. Photoelectric absorption of X-rays by the interstellar matter is the primary factor which limits the distance over which an X-ray can travel. A number of papers have treated the interstellar absorption of cosmic X-rays (see Cruddace et al. (1974) and references cited therein). Brown and Gould (1970) used a procedure similar to that outlined in Section 7.3 except that they used a polynomial fit to the data of Henke, White and Lundberg (1957) to describe the changes in s from threshold to high energies and they included L-shell contributions for oxygen and neon (see also Weisheit

(1974) for a compilation of inner shell photoionization cross sections computed in this manner). The primary uncertainty in the calculations is the assumed abundances of the elements. Brown and Gould assumed: Element ($\log_{10}N$) = H(12.00), He(10.92), C(8.60), N(8.05), O(8.95), Ne(8.00), Mg(7.40), Si(7.50), S(7.35), Ar(6.88).

Figure 7.1 shows a plot of the total effective photoionization cross section per hydrogen atom times a factor $(h\nu/1 \text{ keV})^3$ which has been inserted to flatten the curves, where, for an ion density N_Z and a hydrogen (or proton) density N_H

$$\sigma_{eff} = \sum_Z [N_Z/N_H]\, \sigma_Z \tag{7-45}$$

The photoelectric absorption coefficient of the interstellar medium is then

$$\mu(\nu) = N_H\, \sigma_{eff} \tag{7-46}$$

The edges or discontinuities due to the threshold energies of the various elements are obvious. The energies of these thresholds can be derived from the empirical formula given in Section 7.3. More exact values can be found in House (1969), Wiese, Smith and Glennon (1966)and (1969).

From Figure 7.1 we see that the absorption coefficient decreases as $(h\nu)^{-8/3}$ in the 1-10 keV range and the observed spectrum I will be modified from the intrinsic spectrum I_o after passage through the interstellar medium (see Equation (7-41))

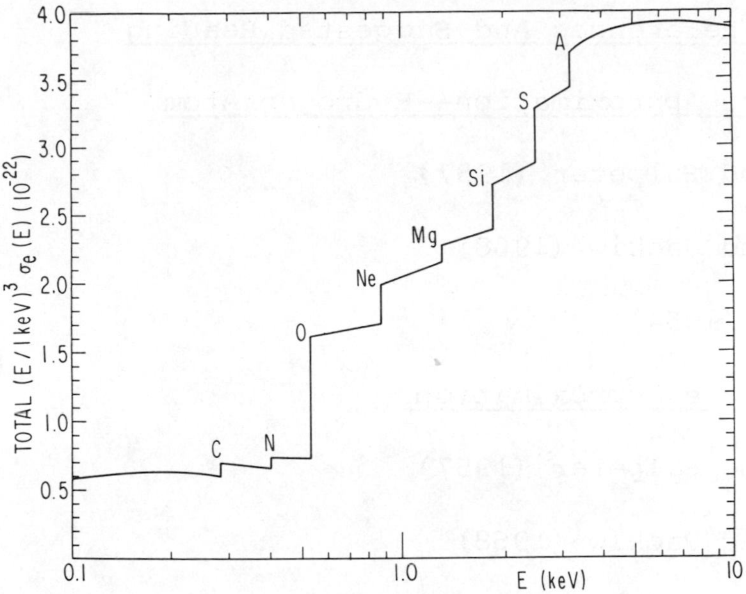

Figure 7.1. Total photo-ionization cross section per hydrogen atom as a function of incident photon energy for a gas having cosmic elemental abundances.

$$I(\nu) \cong I_o(\nu)\ e^{-C/\nu^{8/3}} \tag{7-47}$$

where C is a constant.

Thus the spectrum will turn over at low frequencies. If we assume that the intrinsic spectrum is flat or increasing with decreasing frequencies (a good assumption for many sources) then the shape of the soft X-ray spectrum can be used to estimate the constant C and therefore the number of atoms along the line of sight. If the density is approximately known, we can then estimate the distance to the source, or vice-versa (see Gorenstein (1974) and Problem 7.5).

General References And Suggested Reading

7.1. Born Approximation--Hydrogen Atom

Bethe and Salpeter (1957)

Bethe and Jackiw (1968)

Heitler (1954)

7.2. Dipole Approximation

Bethe and Salpeter (1957)

Bethe and Jackiw (1968)

Frank-Kamenetskii (1962)

7.3. K-Shell Photoionization Cross Section

Bethe and Salpeter (1957)

Daltabruit and Cox (1972)

Brown (1971)--He-like ions

Henke, White and Lundberg (1957)--experimental data

Henke, Elgin, Lent and Ledingham (19)--experimental data

7.4. X-ray Absorption of the Interstellar Medium

Brown and Gould (1970)

Gorenstein (1974)

Cruddace, Paresce, Bower and Lampton (1974)

Other Topics

Photoionization Cross Sections for Complex Atoms and Ions

Theory:

Chapman and Henry (1971, 1972)

Henry and Lipsky (1967)

Bates (1946)

Henry (1970)

Henry and Williams (1968)

Seaton (1958)

Data:

Hudson and Keiffer (1970)

Henke, White and Lundberg (1957)

Weisheit (1974)--semi-empirical fit to data on subshell cross sections for ionization by X-rays.

Problems

7.1. Consider a source of ionizing radiation of luminosity L which is embedded in a uniform cloud of hydrogen of density N which is optically thin to the ionizing photons. Assuming a temperature $T = 10^4$ °K, and a flat spectrum extending out to 1 keV where it cuts off, calculate the radius at which the hydrogen is 50% ionized. How sensitive is the result to the assumed cutoff? How do the results scale with Z?

7.2. Assume that a supernova emits 10^{51} ergs in 20 eV photons in its initial outburst, and that all this energy is deposited within one photo-ionization mean free path in a medium of density 1 hydrogen atom/cm^3. Estimate resulting temperature in the ionized sphere. (See Morrison and Sartori (1969)).

7.3. The intrinsic efficiency of the gas proportional counters commonly used in X-ray astronomy is given by the probability of a photon not being absorbed in the window and then being absorbed in the gas (see Equation (7-41))

$$\text{efficiency} = \exp(-\mu_w R_w)(1 - \exp(-\mu_g R_g))$$

where the subscripts w and g refer to the window and the gas. Compute the efficiency as a function of energy of a counter composed of a window of 4.62 mg/cm^2 Beryllium gas with 3.79 mg/cm^2 Argon (see Giacconi, Gursky and Van Speybroeck (1968)).

7.4. Soft X-ray observations of the Cygnus Loop imply an optical depth of unity at photon energies $\simeq$ 0.30 keV. Assuming a distance of 770 pc, estimate the average density along the line of sight.

7.5. Compute the energy at which the electron scattering cross section becomes equal to the photoelectric cross section for Z = 8 and Z = 26, and for the interstellar medium where $N_8/N_H = 5\times10^{-4}$; $N_{26}/N_H = 3\times10^{-5}$.

Chapter 8

EMISSION AND ABSORPTION LINES

Spectroscopy, or the study of emission and absorption lines has provided us with most of the quantitative knowledge we have of the universe on a scale ranging from stars (spectral typing of stars, determination of stellar masses in binary systems) to the galaxy (21 cm mapping of spiral arms) to the metagalaxy (quasar redshifts, and the expansion of the universe). The fact that an atom or ion of a given element emits and absorbs radiation at a well defined frequency characteristic of the particular element and ionization stage enables us to determine mass motions from the Doppler broadening or shifting of these lines; the intensity of the lines allows us to investigate the amount of matter in a given state of ionization, etc. A book of this type cannot hope to present an adequate discussion of the enormous literature on spectroscopic astrophysics, which includes laboratory and theoretical determinations of atomic parameters, as well as astrophysical applications. Here I give only a sketchy outline which I hope will at least be useful as a guide to reading the literature.

8.1. Selection Rules For Orbital And Magnetic Quantum Numbers--One Electron Spectra

In Section 2.8 we saw that in the dipole approximation ($kr \ll 1$, $v \ll c$, $Z\alpha_f \ll 1$) the probability for a spontaneous transition from a state i to a state f is given by

$$w_{fi} = (32\pi^3/3)(e^2/\hbar c)(\nu^3/c^2)(r)^2_{fi} \qquad (2\text{-}128)$$

where

$$(r)_{fi} = \int u_f^* \, (\hat{\ell}\cdot\vec{r})u_i \; dxdydz \qquad (8\text{-}1)$$

Since the operator $\vec{r}$ which determines the transition probability has odd parity (changes sign under inversion of the spatial coordinates) it follows that the integral (8-1) will vanish when u_f and u_i have the same parity. Hence, transitions between states of the same parity are forbidden (Laporte's rule):

$$(\text{Parity})_i \neq (\text{Parity})_f \qquad (8\text{-}2)$$

For a single electron in a central potential a state with orbital and magnetic quantum numbers $\underline{l}$ and $\underline{m}$ has odd (even) parity if $\underline{l}$ is odd (even) independent of the value of $\underline{m}$. Thus, only those transitions are allowed (in the dipole approximation) which connect a state of odd $\underline{l}$ to one of even $\underline{l}$ (or vice versa).

For an atom with a single electron the eigenfunctions can be expressed in the form

$$u_{n\ell m} = R_{n\ell}(r)\; P_{\ell m}(\cos\theta)e^{im\varphi}C(\ell,m) \qquad (8\text{-}3)$$

The $P_{\ell m}(\cos\theta)$ are the associated Legendre polynomials, and $C(\ell,m)$ is a normalizing constant:

$$C(n,\ell) = ((2\ell+1)(\ell-m)!/4\pi(\ell+m)!)^{\frac{1}{2}} \qquad (8\text{-}4)$$

The associated Legendre polynomials are defined by the equation

$$P_{\ell m}(x) = ((-1)^m/2^\ell \ell!)(1-x^2)^{m/2}\; D_{\ell m}(x)$$

$$D_{\ell m}(x) = (d^{\ell+m}/dx^{\ell+m})(x^2 - 1)^{\ell} \qquad (8\text{-}5)$$

They satisfy the relations

$$P_{\ell,-m}(x) = (-1)^m\ (\ell-m)!\ P_{\ell m}(x)/(\ell+m)! \qquad (8\text{-}6)$$

$$\int_{-1}^{1} P_{\ell' m}(x)\ P_{\ell m}(x)dx = 2/2\ell+1\ \frac{(\ell+m)!}{(\ell-m)!}\ \delta_{\ell'\ell} \qquad (8\text{-}7)$$

Some explicit forms are

$$C(00)P_{00} = (1/4\pi)^{\frac{1}{2}}$$

$$C(11)P_{11} = -(3/8\pi)^{\frac{1}{2}}\ \sin\theta$$

$$C(10) = (3/4\pi)^{\frac{1}{2}}\ \cos\theta \qquad (8\text{-}8)$$

(See Jackson (1962) and references.)
The matrix element of the coordinate z corresponding to a transition from the state having quantum numbers n,l,m to the state n′,l′, m′ (since $z = r\cos\theta$), is

$$(z)_{n'\ell'm',n\ell m} = \int_0^{\infty} r^3\,dr\ R_{n'\ell'}(r)\ R_{n\ell}(r)$$

$$\times \int_0^{\pi} P_{\ell' m'}(\theta)P_{\ell m}(\theta)\cos\theta\sin\theta\,d\theta$$

$$\times \int_0^{2\pi} e^{i(m-m')\varphi}\ d\varphi \qquad (8\text{-}9)$$

If $m' \neq m$, the integral over φ vanishes, so we obtain a selection rule for the magnetic quantum number for radiation emitted with polarization parallel to z:

$$\Delta m = m' - m = 0 \tag{8-10}$$

Using the orthogonality relation (8-7) obeyed by the associated Legendre functions (8-5), we obtain the result that the integral over θ vanishes unless the selection rule for the orbital quantum number is satisfied

$$\Delta \ell = \ell' - \ell = \pm 1 \tag{8-11}$$

Considering matrix elements of the linear combinations

$$x \pm iy = r \sin\theta \; e^{\pm i\varphi} \tag{8-12}$$

One obtains a selection rule for m:

$$\Delta m = m' - m = \pm 1 \tag{8-13}$$

Thus, there can be no radiation polarized parallel to the x and y axes unless $\Delta m = \pm 1$. Classically, this corresponds to the fact that the radiation carries away angular momentum. Evaluating the θ-integrals one again obtains the selection rule (8-11) for the orbital quantum number.

8.2. Selection Rules For Complex Atoms

The above results are valid for a single electron in a central field. In alkali atoms the transitions of most practical importance are those between states in which only the loosely bound valence electron is excited. For such transition, the alkali atom can be treated to a good approximation as a system with only one electron which moves in a central potential and the results stated above still apply.

There are some section rules which apply generally in the dipole approximation.

Consider as arbitrary any electron atom treated by the Russell-Saunders approximation, in which the spin-orbit coupling is small and L and S (the quantum numbers for total orbital and spin angular momentum, respectively) are good quantum numbers. Since $\vec{r}$ commutes with $\vec{S}$, $\vec{r}$ can not connect states with different S or M_s. Hence

$$\Delta S = 0 \tag{8-14}$$

and

$$\Delta M_s = 0 \tag{8-15}$$

The arguments concerning parity changes are still valid, so the parity of the wave function must change in the transition. If we are considering a wave function in the form of products of single-electron wave functions with orbital quantum numbers $\ell_1, \ell_2, \ldots$, then Laporte's rule states that $\sum_k \ell_k$ changes by an odd integer in the transition.

Using general operator methods (see Bethe and Jackiw, (1968), Mizishuma (1968), Condon and Shortley (1963)), it can be shown that

$$\Delta L = 0, \pm 1 \tag{8-16}$$

$$\Delta J = 0, \pm 1 \tag{8-17}$$

The magnetic quantum number M_L remains unchanged if the emitted radiation is polarized parallel to z and changes by unity if the radiation is polarized perpendicular to z:

$$\Delta M_L = 0 \quad \text{(polarized parallel to z)}$$
$$\Delta M_L = \pm 1 \quad \text{(polarized perpendicular to z)} \qquad (8\text{-}18)$$

From symmetry considerations one also finds that transitions between two states with L=0 are forbidden.

$$L = 0 \nrightarrow L = 0 \quad \text{forbidden} \qquad (8\text{-}19)$$

Also,

$$J = 0 \nrightarrow J = 0 \quad \text{forbidden} \qquad (8\text{-}19')$$

For a system of any number of electrons in a central potential, the z components of the total orbital angular momentum (M_L) and of the total spin (M_s) are <u>not</u> constants of motion individually. On the other hand $M = M_L + M_s$ and J are good quantum numbers for any atom in the absence of external fields. The following rigorous selection rules hold:

$$\Delta M = 0, \pm 1 \quad \text{(polar. } \parallel \text{ and } \perp \text{ z, respectively)} \qquad (8\text{-}20)$$

$$\Delta J = 0, \pm 1 \qquad (8\text{-}17)$$

$$J = 0 \nrightarrow J = 0 \quad \text{forbidden} \qquad (8\text{-}19')$$

Parity must change (8-21)

(Condon and Shortley, (1963)).

8.3. Oscillator Strength Sum Rules

Sum rules for oscillator strengths have played an important role in attempts to determine absolute oscillator strengths from sets of relative values, and in checking the internal consistency of calculations or measurements. Here a few sum rules are stated; for proof see Bethe and Salpeter (1957).

The most important sum rule is the Thomas-Reiche-Kuhn rule for all transitions which start from a definite state n of the atom. This is a very general rule which holds for any atom with or without external fields, for any polarization direction and any coupling (L-S or j-j) scheme. Let Z be the total number of electrons in the system and let i be a particular eigenstate of the total Hamiltonian and f any one of a complete set of eigenstates. The sum rule then states that

$$\sum_f f(f,i) = Z \qquad (8\text{-}22)$$

Referring to the definition (2-132) we see that the oscillator strength corresponding to the transition i → f depends on the magnetic quantum members m and m' of the initial and final states. Let us define an average oscillator strength of the transition nl → n′l′ which is independent of polarization and m as follows

$$\overline{f}(f,i) = (2\ell+1)^{-1} \sum_{m'=-\ell'}^{\ell'} \sum_{m=-\ell}^{\ell} f(f,i) \qquad (8\text{-}23)$$

Note that $\bar{f}(f,i)$ is *not* equal to $\bar{f}(i,f)$:

$$\bar{f}(i,f) = (2\ell' + 1)^{-1} \sum_{m'} \sum_{m} f(f,i) \tag{8-24}$$

$$= -(2\ell + 1)\bar{f}(f,i) = -(g_f/g_i)\bar{f}(i,f)$$

where

$$g_f = 2\ell + 1 \tag{8-25}$$

is the degeneracy of the state. The average oscillator strengths obey a sum rule which is stronger than the *f* sum rule:

$$\sum_{n'} \bar{f}(n'\,\ell-1, n\ell) = -\ell(2\ell-1)/3(2\ell+1) \tag{8-26}$$

$$\sum_{n'} \bar{f}(n'\,\ell+1, n\ell) = (\ell+1)(2\ell+3)/3(2\ell+1) \tag{8-27}$$

This is the *Wigner-Kirkwood* sum rule. If these two equations are added, one obtains the *f* sum rule again.

The Wigner-Kirkwood, or "partial f sum rules", show that among the transitions $n\ell \rightarrow n'\,\ell-1$ the ones which lead to energetically lower states (negative ν_{fi}, corresponding to emission) predominate, whereas in the transitions $n\ell \rightarrow n'\,\ell+1$ absorption makes the larger contribution. Absorption predominates also in the summation of all oscillator strengths, according to the ordinary sum rule. Since the energy increases with the principal quantum number, the Wigner-Kirkwood sum rules show that a change of principal and orbital quantum number in the same sense is more probable than a jump in the opposite sense. Thus 2p → 3d is more probable than 2p → 3s.

For any many-electron atom to which the L-S coupling approximation applies, the following sum rule applies:

The total probability of all transitions from a state nLJM to the sub-levels of the state n′L′ (different J′M′) is the same as the total probability for a transition nLM_L to the sublevels of the state n′L′ in the theory without spin.

$$\sum_{j'm'} |\langle n'\ell' j' m' |\vec{r}| n\ell jm\rangle|$$

$$= \sum_{m'_\ell} |\langle n'\ell' m'_\ell |\vec{r}| n\ell m_\ell\rangle|^2 \qquad (8\text{-}28)$$

Hence, the lifetime of any fine structure level is independent of the total quantum numbers J and M and is the same as in the theory without spin. As long as there is no preferred spatial direction, the probability of excitation of a given level is also independent of J and M. Therefore the number of electrons in a given state will depend only on the statistical weight of that state.

This has the consequence that the total intensity of all spectral lines for transitions from a level nLJ (summed over M) to all levels with a fixed n′L′ (all J′M′) is proportional to the statistical weight 2J + 1 of the initial level. For application of this sum rule to the determination of the relative intensities in doublet and triplet spectra, see Bethe and Salpeter (1957). Extensive intensity tables are given in Moore (1959) and Allen (1963).

8.4. Oscillator Strengths And Transition Probabilities For Hydrogen

In order to arrive at the absolute values of the transition probabilities the radial integrals defined in (8-3) must be evaluated:

$$(r)_{n\ell,n'\ell'-1} = \int_0^\infty R_{n\ell}\, R_{n'\ell-1}\, r^3\, dr \qquad (8\text{-}29)$$

For hydrogen the radial eigenfunctions are the associated Laguerre functions (see Bethe and Salpeter, (1957)).

For the Lyman series (1s → np)

$$(r)^2_{n1,10} = (2^8\, n^7 (n-1)^{2n-5}/(n+1)^{2n+5})(a_0/Z)^2 \qquad (8\text{-}30)$$

The mean oscillator strengths for the Lyman series are given by

$$\overline{f}(n1,10) = 2^8 n^5 (n-1)^{2n-4}/3(n+1)^{2n+4} \qquad (8\text{-}31)$$

Since

$$h\nu_{n1}/I_H = (n^2 - 1)/n^2 \qquad (8\text{-}32)$$

The transition probability is given by

$$w_{n1,10} = (8\times 10^9 \times 2^8\, n(n-1)^{2n-2}/3(n+1)^{2n+2})\ \sec^{-1} \qquad (8\text{-}33)$$

provided the ground state is regarded as the initial state. The probability for the radiative transition of an excited np electron to the ground state is obtained by dividing by 3, the statistical weight of the p state (see Equation (8-24).

$$w_{10,nl} = w_{nl,10}/3 \qquad (8\text{-}34)$$

The emitted intensity per np electron is given by

$$P_{10,nl} = h\nu_{nl} w_{10,nl} \qquad (8\text{-}35)$$

(Bethe and Salpeter (1957)(see also Goldwire (1968), Menzel (1969)). Green, Rush and Chandler (1957) give tables of the oscillator strengths, transition probabilities and line intensities for hydrogen. They draw the following general conclusions.

The sum of the oscillator strengths $f(n'\ell' m' \rightarrow n\ell m)$ over all ℓ ℓ' m' and <u>m</u> with fixed <u>n</u> and <u>n'</u> can be approximated accurately for large <u>n</u> and <u>n'</u> (and to within a factor ~ 2 for all values of $n \neq n'$) by

$$f(n', n) = \sum_{\ell \ell' mm'} f(n'\ell' m' \rightarrow n\ell m)$$

$$\simeq (64/3^{3/2}\pi) n^3 \, n'^3 / (n'^2 - n^2)^3 \qquad (8\text{-}36)$$

The transition probabilities are proportional to the frequency so transitions corresponding to high frequencies are most probable, in spite of the fact that the oscillator strengths are largest when ν is smallest. For example, in the transitions from $4p \rightarrow 1s$,

2s and 3s, the ratio of the oscillator strengths is 1/3.5/16 whereas the ratio of the transition probabilities is 23/3/1.

Of all the possible transitions (in emission) from an initial state n, ℓ the transition to the state of lowest energy (compatible with the selection rules) is by far the most probable one, i.e., to the state $n' = 1$, $\ell' = \ell - 1$. The most likely form of a cascade from a state $n\ell$ is the shortest possible one with ℓ steps down to the ground state. Hence, states with $n > \ell+1$ are more easily obtained by direct excitation from the ground state rather than indirectly by excitation or recombination to a higher state followed by a radiative transition or by recombination. Exceptions to this rule are states which can not be excited by a direct radiative transition from the ground state. Such states are obtained indirectly by cascades and directly by electronic excitation.

The lifetimes of the quantum states go up with increasing principal quantum number. For a fixed value of ℓ

$$t(n, \ell) \propto n^3$$

The average lifetime of the nth quantum state is

$$t(n) = \left(\sum_{\ell} (2\ell+1)/n^2 \; (1/t_{n\ell}) \right)^{-1} \propto n^{4.5} \quad (8\text{-}37)$$

8.5. Oscillator Strengths For Complex Atoms And Ions

The various methods for computing oscillator strengths for complex atoms and ions have been reviewed by Layzer and Garstang (1968).

One of the most useful methods for astrophysical purposes has been the Z-expansion approximation. The physical idea underlying this approximation is that, in a first approximation, each electron in a many electron atom moves in a screened Coulomb field. Layzer and Garstang give a detailed discussion of this approximation (see also Cohen, 1967). The general form of the expansion can be roughly understood as follows. According to Equation (2-132) the oscillator strength

$$f(f,i) = (h\nu/I_H)\,(r)^2_{fi}/3\,a_o^2 \qquad (2\text{-}132)$$

Assuming a screened Coulomb field we may write

$$W_i = (Z-S_i)^2 I_H/n_i^2 = (Z^2 I_H/n_i^2)[1- 2S_iZ^{-1} + S_i^2 Z^{-2}]$$

$$W_f = (Z-S_f)^2 I_H/n_f^2 = (Z^2 I_H/n_f^2)[1- 2S_fZ^{-1} + S_f^2 Z^{-2}] \qquad (8\text{-}38)$$

$$(r)^2_{fi} \propto a_o^2 (Z-S)^{-2} = (a_o/Z)^2 (1+2SZ^{-1} + 3S^2 Z^{-2} + \ldots) \qquad (8\text{-}39)$$

Since $h\nu = W_f - W_i$, we have

$$f(f,i) = f_o + f_1 Z^{-1} + f_2 Z^{-2} + \ldots \qquad (8\text{-}40)$$

where f_o is the oscillator strength computed with strictly hydrogenic functions and f_1 is calculated using first order perturbation theory.

The transitions can be grouped into two general categories according to whether or not a change in the principal quantum number occurs. For those transitions in which $n_i \neq n_f$, the lead term f_o is nonzero. For alkali atoms which consist of a single (valence) electron outside one or more closed shells f_o is the hydrogen oscillator strength. For systems with more than one electron outside closed shell it is the hydrogen oscillator strength modified by the usual factor which takes account of the different statistical weights and multiplet structure (see Section 8.3, Shore and Menzel, 1968, Allen, 1963). Unless equivalent electrons are present in the initial or final state, f_o is a quantity which can be calculated in a strictly hydrogenic fashion.

Using the Z-expansion approximation as a guide, Smith and Wiese (1971) have presented graphical and numerical results for a large number of transitions. Using their graphs, it is possible to obtain fairly accurate oscillator strengths for other transitions by extrapolation or interpolation along isoelectronic sequences.

8.6. Higher Multipole Radiation

The Multipole Expansion. For the discrete spectrum of atoms with Z << 137, we can expand the exponential $e^{i\vec{k}\cdot\vec{r}}$ which occurs in the matrix elements for emission and absorption (Equation (2-119))

$$e^{i\vec{k}\cdot\vec{r}} = 1 + i\vec{k}\cdot\vec{r} - (\vec{k}\cdot\vec{r})^2/2 + \ldots \qquad (8\text{-}41)$$

The first term in this expansion leads exactly

to the electric dipole approximation of the previous sections. The higher terms in the expansion lead to terms which are analogous to the types of radiation obtained by a multipole expansion in classical radiation theory.

Consider only the second term of the expansion (8-41). The correction to the dipole matrix element due to this term is (Problem 8.1)

$$D'_{fi} = -(\omega_{fi}/2c)(-im_e\omega_{fi}\ \langle f|xy|i\rangle + \langle f|L_z|i\rangle) \tag{8-42}$$

where L_z is the z-component of the orbital angular momentum operator (Equation (2-52)). This takes no account of the spin of the electron. To do this we must include the term

$$\langle f|\vec{\sigma}\cdot\vec{k} \times \hat{\ell}e^{i\vec{k}\cdot\vec{r}}|i\rangle\ (ie\hbar(2m_e c) \tag{8-43}$$

(see Equation (2-74)). Since it is already a first order term in $\vec{k}\cdot\vec{r}$ we may replace the exponential by unity. Adding this term to the matrix element (8-42) yields, for propagation along the x-axis, and polarization along the y-axis.

$$D^1_{fi} = -(\omega_{fi}/2c)[-i\ m_e\omega_{fi}\ \langle f|xy|i\rangle + \langle f|L_z + 2S_z|i\rangle] \tag{8-44}$$

The spontaneous transition probability is then given by

$$dw_{fi}/d\Omega = (e^2\nu_{fi}/m_e^2 c^3\hbar)\ |D'_{fi}|^2 \tag{8-45}$$

where we have assumed that the electric dipole

matrix element $D_{fi} = 0$. The first term in (8-45) involves a matrix element of the form $e\langle f|xy|i\rangle$ which corresponds to the electric quadrupole moment in the classical case. Hence the radiation due to this term is called <u>electric quadrupole radiation</u>. Introducing the quadrupole moment D from Equation (1-39) and integrating over angles, we can express the transition probability as follows:

$$w^{EQ}_{fi} = (16\pi^5/45)(\nu/c)^5 \hbar^{-1} \ (D)^2_{fi} \qquad (8\text{-}46)$$

The second term involves a matrix element of the form $\langle f|L_z + 2S_z|i\rangle$ $(e/2m_e c)$ which is the quantum mechanical equivalent of the magnetic dipole moment; radiation due to this term is called <u>magnetic dipole radiation</u>. Introducing the magnetic dipole moment $\vec{M}$ (Equations (1-41) and (2-51) and integrating over angles,

$$w^{MD}_{fi} = (32\pi^3/3)(\nu/c)^3 \hbar^{-1} \ (M)^2_{fi} \qquad (8\text{-}47)$$

The selection rules for electric quadrupole radiation of one-electron atoms may be obtained from a consideration of the integration over angles in the matrix element of xy. Another method is to note that, according to the rules of matrix multiplication

$$\langle f|xy|i\rangle = \sum_n \langle f|x|n\rangle \ \langle n|y|i\rangle \qquad (8\text{-}48)$$

Applying the dipole selection rules to the matrix elements of x and y, we obtain

$\Delta \ell = 0, \pm 2$

$\Delta M_\ell = 0, \pm 1, \pm 2$ (Electric Quadrupole)

$\Delta S = 0$ (8-49)

$\Delta M_s = 0$

$\Delta\sigma = 0, \pm 1, \pm 2$

$\ell = 0, \nrightarrow 0$ forbidden

$J = 0 \nrightarrow 0$ forbidden

Parity does not change

For many electron atoms the above rules are esentially unchanged with the substitution of L for l, with the modification that $\Delta L = 1$ is allowed, and $L = 0 \nrightarrow 1$, $L = 1 \nrightarrow 0$ and $J = 0 \nrightarrow 0$ and $J = \frac{1}{2} \nrightarrow \frac{1}{2}$ are forbidden.

In the case of magnetic dipole radiation the relevant matrix element is $\langle f|L+2S|i\rangle = \langle f|J+S|i\rangle$. Since J is a constant of motion, it commutes with the Hamiltonian. Therefore the eigenfunctions of H are also eigenfunctions of J and the relevant matrix element is reduced to $\langle f|S|i\rangle$.

In the case of L-S coupling this matrix element is zero (L, S' are good quantum numbers). The following selection rules apply:

$\Delta \ell_1 = \Delta \ell_2 = \ldots = 0$

$\Delta M = 0, \pm 1$

$\Delta J = 0, \pm 1$; $J = 0 \nrightarrow 0$ forbidden

$\Delta n_1 = \Delta n_2 = \ldots = 0$ (Magnetic Dipole)

Parity unchanged (8-50)

Thus transitions are possible only between states having the same configuration, with same values of L and S.

The spin orbit interaction breaks the L-S coupling by mixing states of different L and S but in general of the same configuration. Since the energy difference of two such states is small, the transition probability is in general low.

Magnetic dipole and electric quadrupole transitions play an important role in astrophysics, in spite of their low transition probabilities relative to dipole transitions. This is because in low density gases an inelastic collision can excite an atom to a metastable level from which a transition is forbidden in the dipole approximation. Thus, if the densities are sufficiently low so that collisional de-excitation is unimportant the atom will decay back to the ground state by a magnetic dipole or electric quadrupole transition producing a "forbidden" line. The forbidden lines of oxygen and the hyperfine 21 cm transition in hydrogen are the most important ones in astrophysical applications. Since they disappear if the density becomes sufficiently high that collisional de-excitation becomes important, their intensities can be used to estimate densities (see Section 8.9).

8.7. Line Profiles

We have seen above that any excited state has a finite lifetime t_i which can be calculated exactly for hydrogenic atoms and ions and approximately for complex atoms and ions (Sections 8.4, 8.5, 2.7, 2.8). According to the uncertainty principle this means that the energy level W_i can only be determined to an accuracy.

$$W_i \simeq \hbar/t_i \tag{8-51}$$

where

$$t_i^{-1} = w_i = \sum_f w_{fi} \tag{8-52}$$

is the probability per second that the state i is vacated through a radiative transition to any other state. Therefore the energy level has a breadth given by (8-51) and the breadth of a line is given by the sum of the breadths of the two levels:

$$W_{fi} \cong \hbar\ (w_i + w_f) \tag{8-53}$$

From this equation it follows that a line may be broad even though the transition probability is very small, provided the final state has a large probability for making a transition to another level (large w_f).

The line profile due to the radiative lifetime may be calculated using time dependent perturbation theory (Section 2.6) and assuming that the initial-final state probability amplitudes have an addition time dependence of $\exp(-wt/2)$. The result is (Heitler, 1954)

$$I(\nu) = I_o\ G(\nu) = I_o(w/2\pi)/((\nu-\nu_0)^2 + (w/2)^2) \tag{8-54}$$

with

$$w = w_i + w_f \tag{8-55}$$

note that

$$\int G(\nu)d\nu = 1 \; ; \; G(\nu_o) = (2/\pi w) \tag{8-56}$$

The line absorption coefficient is given by (see Equations (2-20), (2-118), (2-150))

$$\mu_L(\nu) = (c^2/8\pi\nu^2) A_{fi} G(\nu) N_i F(f,i)$$

$$F(f,i) = (q_f/q_i) - (N_f/N_i) \tag{8-57}$$

Averaging solid angles and introducing the oscillator strength from Equation (2-132)

$$\mu_L(\nu) = \pi r_o c \; f(f,i) N_i \; G(\nu) \; F(f,i) \tag{8-58}$$

The total absorption by the atom is given by an integration over frequency.

$$\mu_L = \pi r_o c \; N_i \; f(f,i) \; F(f,i) \tag{8-59}$$

The full width of the line at half maximum is

$$(\Delta\nu_{if})_{fwhm} = w \simeq 10^9 \; Z^4 \; \text{Hz} \tag{8-60}$$

where the approximate relation is for hydrogenic atoms (see Equation (2-128″). In general, the radiation broadening is much less than the Doppler broadening (see below) near the line center, but in the wings far away from the maximum the radiation broadening dominates. There

$$\mu_L(\nu) = cr_o N_i fwF(f,i)/2\nu^2 \qquad (8\text{-}61)$$

Doppler Broadening. The motion of the atom relative to the observer will introduce a shift in frequency of the line because of the Doppler effect (see Equation (1-80)). For light emitted in the x-direction and non-relativistic velocities the shift amounts to

$$(\nu - \nu_o)/\nu = v_x/c \qquad (8\text{-}62)$$

where v_x is the velocity of the atom relative to the observer (positive velocities in the direction of the observer). For a Maxwellian distribution of atom velocities the number of atoms with a velocity between v_x and $v_x + dv_x$ is

$$f(v_x) = (M/2\pi KT)^{\frac{1}{2}}\, e^{-Mv^2/2KT} \qquad (8\text{-}63)$$

(M = mass of atom or ion).
The corresponding intensity distribution is obtained by substitution of Equation (8-62) into (8-63):

$$G(\nu) = \pi^{-\frac{1}{2}}(\Delta\nu_D)^{-1}\ \exp(-(\nu-\nu_o)^2/(\Delta\nu_D)^2) \qquad (8\text{-}64)$$

where

$$\Delta\nu_D/\nu = (2KT/Mc^2)^{\frac{1}{2}} = 4.3\times10^{-7}\ T^{\frac{1}{2}}\ A^{-\frac{1}{2}} \qquad (8\text{-}65)$$

(A is the atomic weight)
Note that in the case of Doppler broadening, the line width is proportional to the frequency, in contrast to radiation broadening.

The full width at half maximum is

$$(\Delta\nu)_{fwhm} = (\Delta\nu_D) \log_e 2 \qquad (8\text{-}66)$$

Comparison of Equations (8-66) and (8-60) shows that Doppler broadening will dominate at temperatures T greater than about Z^5 for hydrogenic ions. Since T is on the order of 10^4-10^5 Z^2, for almost any excitation conditions, it follows that Doppler broadening · will dominate radiation broadening near the line center for all Z less than about 50.

The absorption coefficient at line center

$$\mu_L(\nu_o) = \pi^{\frac{1}{2}} r_o c \; f(f,i) N_i F(f,i)/\Delta\nu_D \qquad (8\text{-}67)$$

For frequencies for away from the line center, the Doppler broadening decreases exponentially, so there the line profile is determined by radiation broadening (8-61) or possibly collisional broadening.

The large cross section at line center shows that many gas astrophysical plasma may be optically thick to resonance scattering in certain lines, especially Ly. This problem has been discussed in detail by Hummer (1968) (Hummer and Rybicki, 1971).

In addition to thermal motions, mass motions can cause Doppler broadening. Examples are the expansion of an optically thin cloud such as a supernova remnant, or turbulent motions in the source. To compute the exact profile some assumption about the distribution of particle velocities would be needed. In most cases, however, it is sufficient to simply replace the thermal velocity by the characteristic velocity of the mass motions in Equation (8-65).

Collisional Broadening. Collisions can also shorten the lifetime of an excited state by inducing a transition to another level. The effect is analogous to radiation broadening, except that the radiative lifetime is replaced by the time between collisions t_{coll}:

$$\Delta\nu_{coll} \simeq 1/t_{coll} \tag{8-68}$$

The intensity distribution is the same as for radiation broadening (8-54). Collisional broadening becomes important when the collision rate is on the order of 10^9/sec, i.e., for densities greater than about $10^{14}\ T^{\frac{1}{2}}$ °K. Densities of this order are achieved in stellar photospheres.

Compton Scattering. Finally, as discussed in Section 4.6, Compton scattering also broadens lines. The order of magnitude is given by Equation (4-63). The details of the line profile are discussed in Chandrasekhar (1960).

8.8. Emission And Absorption Lines

Whether or not a given line appears in emission or absorption depends on the relative intensities in the line and in the continuum. The intensity in the line is described by the equation of transfer (see Equation (2-148)):

$$I_L = I(0)e^{-\tau(0,s)} + \int_0^s (j_L + j_c)e^{-\tau(s',s)}ds' \tag{2-148}$$

where I(0) is the background intensity at the

line frequency and the optical depth is composed of continuum and line absorption coefficients:

$$\tau = \tau_C + \tau_L \tag{8-69}$$

The equation of transfer for the continuum adjacent to the line is

$$I_C = I(0)e^{-\tau_C(0,s)} + \int_0^s j_C e^{-\tau_C(s,s')} es' \tag{8-70}$$

The intensity of the line relative to the continuum is

$$I_L - I_C = I(0)(e^{-\tau_L(0,s)} - e^{-\tau_C(0,s)})$$

$$+ \int_0^s j_L e^{-\tau(s,s')} ds' \tag{8-71}$$

$$- \int_0^s j_C(e^{-\tau_C(s,s')} - e^{-\tau(s,s')})ds'$$

For almost every case the line absorption coefficient is much greater than the continuous absorption coefficient (cf. Equations (8-67), (5-66), (7-25)). Then Equation (8-71) can be simplified by setting $\tau_C = 0$:

$$I_L - I_C = - I(0)(1- e^{-\tau_L}) + S(0,s)$$

$$S(0,s) = \int_o^s j_L e^{-\tau_L} ds' - \int_o^s j_c (1-e^{-\tau_L}) ds' \quad (8\text{-}72)$$

For a uniform medium in which the levels are populated according to thermal equilibrium (see Equation (2-151))

$$I_L - I_c = -I(0)(1- e^{-\tau_L}) + B_L(\nu)(1- e^{-\tau_L})$$

$$- I_c(1- e^{-\overline{\tau}_L}) \quad (8\text{-}73)$$

$B(T_L)$ is the Planck function for equilibrium radiation, evaluated for a temperature characteristic of the excitation conditions for the line (see Equations (2-21) and (2-151)), and $\overline{\tau}_L$ is an average line absorption optical depth, defined by

$$e^{-\overline{\tau}_L} = \int_o^s j_c\, e^{-\tau_L(s',s)} ds' / \int_o^s j_c\, ds' \quad (8\text{-}74)$$

Departures from thermal equilibrium can be expressed in terms of an excitation temperature T_{ex} (see Equation (2-150), (2-151), (2-152)):

$$e^{h\nu/KT_{ex}} = q_i N_f / q_f N_i \quad (8\text{-}75)$$

Since $\overline{\tau}_L$ and τ_L are of the same order, Equation (8-73) shows that the lines will appear in emission if

$$B(T_L) > I(0) + I_c \qquad \text{(emission)} \qquad (8\text{-}76)$$

On the other hand, if there is a strong background source as is the case for matter just above the photosphere of stars, then the line will appear in absorption:

$$B(T_L) < I(0) + I_c \qquad \text{(absorption)} \qquad (8\text{-}77)$$

In the limit of an optically thin gas with no background

$$I_L = \int j_L \, ds' + \int j_c \, ds' \qquad (8\text{-}78)$$

8.9. Collisional Excitation And Level Populations

In many astrophysical problems the relative populations of the excited states are not described by the thermal Boltzmann distribution; they must be determined by a steady state balance of the microprocesses. For the simple case of a two level atom this balance can be written as (N= number of electrons/cm^3)

$$N_m (w_{mn} + \overline{IB}_{mn} + NC_{mn}) = N_n (\overline{IB}_{nm} + NC_{nm}) \qquad (8\text{-}79)$$

or

$$N_m/N_n = (\overline{IB}_{nm} + NC_{nm})/(w_{mn} + \overline{IB}_{mn} + NC_{mn}) \qquad (8\text{-}79')$$

where $\overline{IB}$ denotes an average over all directions, and the C's are collision rate constants which are independent of particle and radiation densities. In general, electron collisions dominate heavy particle collisions, except for transitions in which the energy change is smaller than the mean thermal energy of the particles (hyperfine and fine structure levels, see Dalgarno and Reid, 1969, and references cited therein).

Electron Collisional Excitation. The classical treatment of excitation of atoms or ions can be carried out in a manner analogous to the treatment of ionization. From Equation (5-2) we see that the cross section for an energy transfer W by an electron of energy W is

$$\sigma \sim \pi b^2 \sim \pi e^4/W\Delta W = 4\pi\, a_o^2 (I_H/W)(I_H/\Delta W) \quad (8\text{-}80)$$

In a semiclassical theory, the impinging electron is treated as a classical charged particle that produces a variable field in the neighborhood of the atom or ion, and quantum perturbation theory is used to calculate the probability that the variable field induces a transition to an excited state. If $K_{nm}(b)$ is the probability that an electron with an impact parameter b will induce a transition from a state n to a state m, the cross section for the n-m transition is

$$\sigma(n,m) = \int_o^\infty K_{nm}(b)\, 2\pi b\, db \quad (8\text{-}81)$$

where, from the definition of $K_{nm}(b)$ as a probability, we must have

$$\sum_{m\neq n} K_{nm}(b) \leq 1 \quad \text{for any b .} \quad (8\text{-}82)$$

We can introduce some quantum theory by quantizing the angular momentum of the electron

$$L_e = m_e vb = \hbar(\ell_e(\ell_e + 1))^{1/2} \quad (8\text{-}83)$$

Then

$$2b\,db = (2\ell_e + 1)(\hbar/m_e v)^2\ \Delta\ell_e \quad (8\text{-}84)$$

Using the fact that $\Delta\ell_e = 1$, we can replace the integral in Equation (8-81) by a sum to obtain

$$\sigma(n,m) = \pi\ a_o^2 (I_H/W) \sum_{\ell_e} (2\ell_e + 1) K_{nm}(\ell_e) \quad (8\text{-}85)$$

Introducing the partial and total collision strengths

$$\Omega(\ell_e, n, m) = (2\ell_e + 1)\ K_{nm}(\ell_e)$$

$$\Omega(n,m) = \sum \Omega(\ell_e, n, m) \quad (8\text{-}86)$$

and averaging over initial states we obtain

$$\sigma(n,m) = (\pi\ a_o^2/q_n)(I_H/W)\Omega(n,m) \quad (8\text{-}87)$$

(q_n = statistical weight of state n)

The condition (8-82) requires that

$$\sum_{m \neq n} \Omega(\ell_e, n, m)/q_n \leq 2\ell_e + 1 \quad (8\text{-}88)$$

which is analogous to the oscillator strength sum rules. Equation (8-88) is useful for obtaining upper limits on the partial cross section.

The evaluation of the collision strength for electrons and ions has been reviewed by a number of authors (see Bely and Van Regemorter, 1970, Czyzak, 1968, and references cited therein). Comparison with the rough classical estimate (8-80) shows that

$$\Omega(n,m)/q_n \sim 4\ I_H/\Delta W \quad ; \quad \Delta W = W_m - W_n \qquad (8\text{-}89)$$

Another estimate which is very similar to (8-89) can be obtained from the Born approximation calculations, valid at high energies (Bethe and Jackiw, 1968, Van Regemorter, 1962)

$$\Omega(n,m)/q_n \sim (8\pi/\sqrt{3})\, f\bar{g}\,(I_H/\Delta W) \qquad (8\text{-}90)$$

where $\bar{g} \simeq 0.2$ for low energies and $f(m,n)$ is the oscillator strength. This is also called the Bethe approximation, and the $\bar{g}$ approximation. The approximation (8-89) and (8-90) are fairly accurate for hydrogenic ions where the principal quantum number changes, but in general they should only be used as a guide to interpolation when exact calculations are not available.

The cross sections for excitation are finite at threshold for ions but vanish for neutral atoms. This is because for ions, even an electron which has very low velocities at large distances from the ion can be accelerated to energies above threshold by the attractive Coulomb field of the ion.

The rate coefficient for collisional excitation is

$$C_{nm} = \int_{v_o}^{\infty} v\ f(v)\sigma(n,m)dv \tag{8-91}$$

where v_o is the threshold velocity

$$v_o = (2\Delta W/m_e)^{\frac{1}{2}} \qquad \Delta W = W_m - W_n \tag{8-92}$$

For a Maxwellian distribution of electron velocities

$$C_{nm} = (\pi a_0^2/q_n)(2KT/\pi m_e)^{\frac{1}{2}}(2I_H/KT)\overline{\Omega}(n,m)$$

$$\cdot\ e^{-\Delta W/KT} \tag{8-93}$$

$$= 8.6\times10^{-6}\ q_n^{-1}\ T^{-\frac{1}{2}}\ \overline{\Omega}(n,m)e^{-\Delta W/KT} \quad \text{cm}^3/\text{sec}$$

where

$$\overline{\Omega}(n,m) = \int_{y_o}^{\infty} \Omega(n,m)e^{-y}\ dy/e^{-y_o} \tag{8-94}$$

$$y_o = \Delta W/KT$$

In thermal equilibrium the levels n and m will be described by a Boltzmann distribution and the rates of excitation C_{nm} and de-excitation C_{mn} must be the same (cf. Equations (2-9), (2-10)). This requires that

$$\Omega(n,m) = \Omega(m,n) \tag{8-95}$$

and

$$C_{nm} = C_{mn}\ (q_m/q_n)e^{-\Delta W/KT} \tag{8-96}$$

Level Populations. We return now to a discussion of the level populations. In the case where background radiation field is low, (8-79′) becomes

$$N_m/N_n = N_e C_{nm}/(w_{mn} + N_e C_{mn}) \quad (8\text{-}97)$$

In the limit of high densities $N_e C_{mn}$ is much greater than w_{mn} and the level populations are described by the Boltzmann distribution (2-10). In the limit of low densities

$$N_m/N_n = N_e \; C_{nm}/w_{mn} \quad (8\text{-}98)$$

and the power emitted per atom (or ion) in the ground state is

$$P_{mn} = N_e \; C_{nm} \; h\nu_{mn} \; ; \; h\nu_{mn} = W_m - W_n \quad (8\text{-}99)$$

For low densities where (8-98) holds, almost all of the atoms will be in the ground state, so the power emitted per unit volume in a line ν_{mn} from an atom or ion z, z is given by

$$P_{mn} = N_e N_{Z,z} 8.6\times10^{-6} \; T^{-\frac{1}{2}} \; (h\nu)_{mn}$$

$$\cdot \; (\Omega(m,n)/q_n) e^{-(h\nu)_{mn}/KT} \quad (8\text{-}100)$$

where $N_{Z,z}$ is the number of atoms or ions of element Z and ionization stage z in the gas. Its value depends on the abundances of the elements and the ionization state of the gas.

The latter is determined by balancing the rates of ionization and recombination. In the limit where photoionization is unimportant the ionization by electron collisions is balanced by radiative and dielectronic recombination. Since the ionization and recombination rates by these processes are all proportional to the electron density, the calculated ionization equilibria are independent of the electron density:

$$n_{Z-1}/n_Z = \alpha_Z/c_{Z-1} \tag{8-101}$$

where c_Z is the collisional ionization rate coefficient.

For a calculation of the low density ionization equilibria for a number of ions, see Cox and Tucker (1969), Jordan (1969, 1970), Beigman, Vainshtein and Vinogradov (1970).

For low densities, line emission is the most important radiative energy loss mechanism for a plasma having a temperature less than about ten million degrees, in spite of the fact that the principal contributors at temperatures above about 30,000 degrees are elements whose abundance is down by a factor of a thousand from the abundance of the protons and electrons which determine the bremsstrahlung and recombination emission. This is due to the much larger cross section for collisional excitation as compared with bremsstrahlung and radiative recombination (cf. Equations (8-87), (5-28), (6-4)). Figure 8.1 (Cox and Tucker, 1969) show the total radiative power due to bremsstrahlung, recombination and line radiation as a function of temperature for a gas with abundances $A(\log_{10} N)$ = H(12.00), He(11.20), C(8.60), N(8.04), O(8.95), Ne(8.70), Mg(7.43), Si(7.50) and S(7.30) and an ionization balance computed from (8-101). The dotted curve was computed on the assumption that dielectronic recombination was unimportant (see Chapter 6). Although the details of the ionization balance may significantly change the power emitted by a given line, they clearly do not change the general features of the total radiative cooling curve which is determined by a large number of ions. The principal contributors to the radiative energy loss are denoted. It shows that bremsstrahlung (B) is significant only for temperatures somewhat greater than a million degrees.

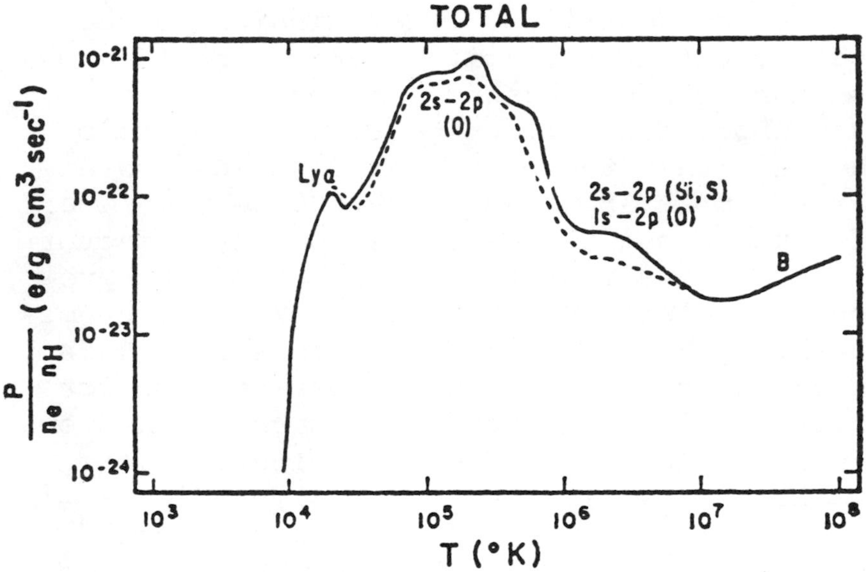

Figure 8.1. Total radiative power loss due to bremsstrahlung, recombination radiation and line emission for a function of temperature for a low density gas with the cosmic abundances.

General References And Suggested Reading

8.0. General References

The following books and review articles are good general references for the material discussed in this chapter:

Bethe and Salpeter (1957)

Bethe and Jackiw (1968)

Condon and Shortley (1963)

Czyzak (1968)

Gebbie (1971)

Griem (1974)

Mizushima (1970)

Shore and Menzel (1968)

Other References for Specific Topics:

8.1., 8.2., and 8.3. Selection Rules and Sum Rules

Bethe and Jackiw (1968)

Bethe and Salpeter (1957)

Condon and Shortley (1963)

Mizishuma (1970)

Shore and Menzel (1968)

8.4. Oscillator Strengths and Transition Probabilities for Hydrogen

Bethe and Salpeter (1957)

Goldwire (1968)

Menzel (1969)

Green, Rush and Chandler (1957)

8.5. Oscillator Strengths for Complex Atoms and Ions

Brown (1972)--He-like

Cohen and Kelley (1967)--He-like

Drake and Dalgarno (1971)--He-like

Griem (1963)

Layzer and Garstang (1968)

Shore and Menzel (1968)

Smith and Wiese (1971)

8.6. Higher Multipole Radiation

Dalgarno (1971)

Layzer and Garstang (1968)

Osterbrock (1964)

Seaton (1960)

Any of the references given under 8.0.

8.7. Line Profiles

Heitler (1954)

Jeffries (1968)

Griem (1974)

Hummer and Rybicki (1971)

Van Regemorter (1965)

8.8. Emission and Absorption Lines

Hummer and Rybicki (1971)

Jeffries (1968)

Pacholczyk (1970)

8.9. Collisional Excitation and Level Populations

Bely (1970)

Bethe and Jackiw (1968)

Burgess, Hummer and Tully (1970)

Czyzak (1968)

Sampson (1974)

Seaton (1962)

8.10. Other Topics

21 cm Radiation:

Kerr (1968)

Pacholczyk (1970)

Shklovsky (1960)

Molecular Lines:

Zuckerman and Palmer (1974)

Field, Sommerville and Dressler (1966)

Litvak (1974)

Rank, Townes and Welch (1971)

Robinson and McGee (1967)

Solomon (1973)

Two Photon Decay:

Drake, Victor and Dalgarno (1969)

Problems

8.1. Show that the following operator relation is valid:

$$H(xy) - (xy)\,H = (\hbar/m_e i)(p_x y + x p_y)$$

and consequently that

$$(p_x y + y p_x)_{fi} = i m_e \omega_{fi}$$

Use the above identity to derive Equation (8-42)(Feynman, 1962).

8.2. For the 21 cm hydrogen line the Einstein transition probability is $A_{10} = 2.85\times10^{-15}\ \mathrm{sec}^{-1}$ and the statistical weights of the upper and lower levels are $q_1 = 3$, and $q_o = 1$.

(a) Use Equation (8-57) to compute the line absorption coefficient.

(b) Expressing the intensity in terms of the brightness temperature, T_b, show that, for an optically thin medium the number of ground-level atoms in unit frequency interval in a cylinder of cross section 1 cm^2 extending along the entire line of sight is

$$N_H(\nu) \simeq 3.9\times10^{14}\ T_B(\nu)$$

or, using Equation (8-62), that

$$N_H(v) \simeq 1.8\times10^{18}\ T_B(v)$$

where $N_H(v)$ is the number of atoms in a

velocity interval of 1 km/sec. The total number of atoms in the line of sight is obtained by integrating over the line profile. (Kerr, 1968, Pacholczyk, 1970, Shklovsky, 1960).

8.3. Use Equation (8-73) to show that the optical depth of an absorption line can be expressed as

$$\tau_L = - \log(1 - (\Delta T_{off} - \Delta T_{on})/T_{bkgd})$$

where T_{bkgd} is the brightness temperature of the background source, ΔT_{off} is the difference in the brightness temperatures in the line and in the adjacent continuum when looking away from the background source and ΔT_{on} is the same quantity obtained when looking at the source.

8.4. The collisional ionization rate for a hydrogenic ion is approximately

$$c_Z \simeq 10^{-10} T^{\frac{1}{2}} Z^{-4} \exp(-Z^2 I_H/KT)$$

Use this expression together with Equations (6-9) and (8-101) to compute the ionization equilibrium of a hydrogenic plasma as a function of temperature. Compare the result with the Saha formula (problem 2.8).

8.5. Derive Equation (6-34), using the results of this chapter and problem 2.8.

8.6. Compute the absorption coefficient of the Lyman alpha line on hydrogen (a) in the Doppler core; (b) in the wings. Compute the mean free path for absorption of Lyman alpha in the interstellar medium for a hydrogen density of 1 particle per cm^3.

Aldrovandi, S. and Pequignot, D. (1973) Astr. Ap., 25, 137.

Alfven, H. and Falthammar, C-G. (1963) Cosmical Electrodynamics (London: Oxford University Press).

Allen, C. (1963) Astrophysical Quantities (London: Athlone Press).

Aller, L. (1961) The Abundances of the Elements (New York: Interscience).

Aller, L. and Liller, W. (1968) in Nebulae and Interstellar Matter, ed. B. Middlehurst and L. Aller (Chicago: University of Chicago Press).

Bates, D. (1946) M.N.R.A.S. 106, 432.

Bates, D. and Damgaard, A. (1949) Phil. Trans. Roy. Soc. (London), 242, 101.

Beigman, I., Vainshtein, L. and Vinogradov, A. (1970) Sov. Astr. 13, 775.

Bekefi, G. (1966) Radiative Processes in Plasmas (New York: Wiley).

Bely, O. and Van Regemorter, H. (1970) Ann. Rev. Astr. Ap. 8, 329.

Bethe, H. and Ashkin, J. (1953) Experimental Nuclear Physics, Vol. 1, ed. E. Segre (New York: Wiley).

Bethe, H. and Jackiw, R. (1968) Intermediate Quantum Mechanics (New York: Benjamin).

Bethe, H. and Salpeter, E. (1957) Quantum Mechanics of One and Two Electron Atoms (New York: Academic Press).

Blokhintsev, D. (1964) Quantum Mechanics (Dordrecht: Reidel).

Blumenthal, G. and Gould, R. (1970) Rev. Mod. Phys., 42, 237.

Brown, R.L. and Gould, R. (1970) Phys. Rev. D2, 2252.

Brown, R.T. (1971) Ap. J. 170, 387.

Burbidge, G. (1959) in Paris Symposium on Radio Astronomy, ed. R. Bracewell (Stanford: Stanford University Press).

Burbidge, G., Gould, R. and Pottasch, S. (1963) Ap. J. 138, 945.

Burgess, A. (1964) Ap. J. 139, 776.

Burgess, A. (1965) Ap. J. 141, 1588.

Burgess, A. and Summers, H. (1969) Ap. J. 157, 1007.

Burgess, A., Hummer, D. and Tully, J. (1970) Phil. Trans. Roy. Soc. London, 266, 225.

Burn, B. (1966) M.N.R.A.S., 133, 67.

Canuto, V., Lodenquai, J. and Ruderman, M. (1971) Phys. Rev. D3, 2303.

Chandrasekhar, S. (1960) Radiative Transfer (New York: Dover).

Chapman, R. and Henry, R. (1971) Ap. J. 168, 169.

Chapman, R. and Henry, R. (1972) Ap. J. 173, 243.

Cohen, M. (1967) Canad. J. Phys. 45, 2009.

Cohen, M. and Kelley, P. (1967) Canad. J. Phys. 45, 2079.

Condon, E. and Shortley, G. (1963) The Theory of Atomic Spectra (Cambridge: Cambrdige University Press).

Cox, D. and Tucker, W. (1969) Ap. J. 157, 1157.

Cruddace, R., Paresce, F., Bowyer, S. and Lampton, M. (1974) Ap. J. 187, 497.

Culhane, J. (1969) M.N.R.A.S. 144, 375.

Czyzak, S. (1968) in Nebulae and Interstellar Matter, ed. B. Middlehurst and L. Aller (Chicago: University of Chicago Press).

Dalgarno, A. and Reid, R. (1969) Mem. Soc. Roy. Sci. 17, 69.

Dalgarno, A. (1971) in The Menzel Symposium, NBS Special Publication 353 (Washington: Government Printing Office).

Daltabuit, E. and Cox, D. (1972) Ap. J. 177, 855.

Delmer, T., Gould, R. and Ramsay, W. (1967) Ap. J. 149, 495.

Drake, G. and Dalgarno, A. (1971) Proc. Roy. Soc. London A, 320, 549.

Drake, G., Victor, G. and Dalgarno, A. (1969) Phys. Rev. 180, 25.

Dupree, A. and Goldberg, L. (1970) Ann. Rev. Astr. Ap. 8, 231.

Eddington, A. (1926) The Internal Constitution of Stars (Cambridge: Cambridge University Press).

Elwert, G. (1939) Ann. Physik 34, 178.

Elwert, G. (1948) Z. Naturforsch 3a, 477.

Elwert, G. (1954) Z. Naturforsch 9a, 637.

Epstein, R. (1973) Ap. J. 183, 593.

Felten, J. and Morrison, P. (1966) Ap. J. 146, 686.

Felten, J., Adams, T. and Rees, M. (1972) Astron. Ap. 21, 139.

Felten, J. and Rees, M. (1972) Astron. Ap. 17, 226.

Feynman, R. (1962) Quantum Electrodynamics (New York: Benjamin).

Field, G., Sommerville, W. and Dressler, K. (1966) Ann. Rev. Astr. Ap. 4, 207.

Frank-Kamenetskii, D. (1962) Physical Processes in Stellar Interiors (Jerusalem: Israel Program for Scientific Translations).

Gardner, F. and Whiteoak, J. (1966) Ann. Rev. Astron. Ap. 4, 245.

Gebbie, K. (1971), ed. The Menzel Symposium, NBS Special Pub. 353 (Washington, U.S. Government Printing Office).

Giacconi, R., Gursky, H. and van Speybroeck, L. (1968) Ann. Rev. Astr. Ap. 6, 373.

Giacconi, R. and Gursky, H. (1974) X-Ray Astronomy (Dordrecht: Reidel).

Ginzburg, V. (1964) The Propagation of Electromagnetic Waves in Plasmas (London: Pergamon).

Ginzburg, V. and Syrovatskii, S. (1964) The Origin of Cosmic Rays (London: Pergamon).

Ginzburg, V. and Syrovatskii, S. (1965) Ann. Rev. Astr. Ap. 3, 297.

Ginzburg, V. and Syrovatskii, S. (1969) Ann. Rev. Astr. Ap. 7, 375.

Glasco, H. and Zirin, H. (1964) Ap. J. Suppl. 9, 193.

Gnedin, Y. and Sunyaev, R. (1973) M.N.R.A.S. 162, 53.

Goldsmith, D. and Silk, J. (1972) Ap. J. 172, 563.

Goldwire, H. (1968) Ap. J. Suppl. 17, 1171.

Gorenstein, P. (1974) X-Ray Astronomy, ed. R. Giacconi and H. Gursky (Dordrecht: Reidel).

Gorenstein, P., Gursky, H. and Garmire, G. (1968) Ap. J. 153, 885.

Gould, R. (1969) Austral. J. Phys. 22, 189.

Gould, R. (1970) Am. J. Phys. 38, 189.

Gould, R. (1971) Astrophys. L. 8, 129.

Gould, R. (1972) Ann. Phys. 69, 321.

Gould, R. (1972a) Physica 60, 145.

Gould, R. (1972b) Physica 62, 555.

Gould, R. (1975) Ann. Phys. (in press).

Gould, R. and Thakur, R. (1970) Ann. Phys. 61, 351.

Griem, H. (1974) Spectral Line Broadening by Plasmas (New York: Academic Press).

Green, L., Rush, P. and Chandler, C. (1957) Ap. J. Suppl. 3, 37.

Heitler, W. (1954) The Quantum Theory of Radiation (London: Oxford University Press).

Henke, B., White, R. and Lundberg, B. (1957) J. App. Phys. 28, 98.

Henke, B., Elgin, R., Lent, R. and Ledingham, R. (1967) Norelco Reporter 14, Nos. 3 and 4, July-December.

Henry, R. (1970) Ap. J. 161, 1153.

Henry, R. and Lipsky, L. (1967) Phys. Rev. 153, 51.

Henry, R. and Williams, R. (1968) P.A.S.P., 80, 669.

Hewish, A. (1970) Ann. Rev. Astr. Ap. 8, 265.

Hjellming, R. and Davies, R. (1970) Astr. Ap. 5, 53.

House, L. (1969) Ap. J. Suppl. 18, 21.

Houston, W. (1959) Principles of Atomic Spectra (New York: Dover).

Hudson, R. and Kieffer, L. (1970) Bibliography of Photoionization Cross Section Data (JILA Information Center, Report 11).

Hummer, D. (1968) M.N.R.A.S. 138, 73.

Hummer, D. and Rybicki, G. (1970) Ann. Rev. Astr. Ap. 9, 237.

Illarionov, A. and Sunyaev, R. (1972) Sov. Astr. 16, 45.

Jackson, J. (1962) Classical Electrodynamics (New York: Wiley).

Jeffries, J. (1968) Spectral Line Formation (Waltham, Mass.: Blaisdell).

Jones, F. (1968) Phys. Rev. 167, 1159.

Jordan, C. (1969) M.N.R.A.S. 142, 499.

Jordan, C. (1970) M.N.R.A.S. 149, 1.

Kardashev, N. (1959) Sov. Astr. 3, 813.

Kardashev, N. (1962) Sov. Astr. 6, 317.

Karzas, W. and Latter, R. (1961) Ap. J. Suppl. 6, 167.

Kellerman, K. (1966) Ap. J. 146, 621.

Kellerman, K. and Pauliny-Toth, I. (1968) Ann. Rev. Astr. Ap. 6, 417.

Kerr, F. (1968) in Nebulae and Interstellar Matter, ed. B. Middlehurst and L. Aller (Chicago: University of Chicago Press).

Kramers, H. (1958) Quantum Mechanics (Amsterdam: North Holland).

Landau, L. and Lifshitz, I. (1960) Electrodynamics of Continuous Media (Reading, Mass.: Addison Wesley).

Landau, L. and Lifshitz, I. (1962) The Classical Theory of Fields (Reading, Mass.: Addison Wesley).

Layzer, D. and Garstang, R. (1968) Ann. Rev. Astr. Ap. 6, 449.

Legg, M. and Westfold, K. (1968) Ap. J. 154, 499.

Litvak, M. (1974) Ann. Rev. Astr. Ap. 12, 97.

Mathews, W. and O'Dell, C. (1969) Ann. Rev. Astr. Ap. 7, 67.

McGuire, E. (1969) Phys. Rev., 185, 1.

Melrose, D. (1969) Astrophys. Space Sci. 5, 131.

Menzel, D. (1969) Ap. J. Suppl. 18, 221.

Mizushima, (1970) Quantum Mechanics of Atomic Spectra and Atomic Structure (New York: Benjamin).

Moore, C. (1959) A Multiplet Table of Astrophysical Interest (Washington: United States Government Printing Office).

Morris, D. and Berge, G. (1964) Ap. J. 139, 1388.

Morrison, P. and Sartori, L. (1969) Ap. J. 158, 541.

Novikov, I. and Thorne, K. (1973) in Black Holes, ed. C. DeWitt (New York: Gordon and Breach).

O'Dell, S. and Sartori, L. (1970) Ap. J. 162, L37.

Osterbrock, D. (1964) Ann. Rev. Astr. Ap. 2, 95.

Pacholczyk, A. (1970) Radio Astrophysics, San Francisco: Freeman).

Pengelly, R. (1964) M.N.R.A.S. 127, 145.

Petrosian, V. (1973) Ap. J. 186, 291.

Rank, D., Townes, C. and Welch, W. (1971) Science 174, 1083.

Robinson, B. and McGee, R. (1967) Ann. Rev. Astr. Ap. 5, 183.

Rossi, B. (1952) High Energy Particles (Englewood Cliffs, N.J.: Prentice Hall).

Ruderman, M. (1972) Ann. Rev. Astr. Ap. 10, 427.

Sampson, D. (1974) Ap. J. Suppl. 28, No. 263.

Scheuer, P. (1960) M.N.R.A.S. 120, 231.

Scheuer, P. and Williams, P. (1968) Ann. Rev. Astr. Ap. 6, 321.

Schott, G. (1912) Electromagnetic Radiation (Cambridge: Cambridge University Press).

Seaton, M. (1958) Rev. Mod. Phys. 30, 979.

Seaton, M. (1959) M.N.R.A.S. 119, 81.

Seaton, M. (1960) Rep. Prog. Phys. 23, 313.

Seaton, M. (1962) in Atomic and Molecular Processes, ed. D. Bates (New York: Interscience)

Shklovsky, I. (1960) Cosmic Radio Waves (Cambridge: Harvard University Press).

Shore, B. (1969) Ap. J. 158, 1205.

Shore, B. and Menzel, D. (1968) Principles of Atomic Spectra (New York: Wiley).

Silk, J. (1973) Ann. Rev. Astr. Ap. 11, 269.

Smith, M. and Wiese, W. (1971) Ap. J. Suppl. 23, 103.

Solomon, P. (1973) Phys. Today. March issue, p. 32.

Sommerfeld, A. (1939) Atombau and Spektrallinien, Bd. II (Braunschweig: Vieweg).

Spitzer, Jr., L. (1948) Ap. J. 107, 6.

Tarter, C. (1971) Ap. J. 168, 313.

Tarter, C. (1973) Ap. J. 181, 607.

Tsytovich, V. (1973) Ann. Rev. Astr. Ap. 11, 363.

Tucker, W. and Gould, R. (1966) Ap. J. 144, 244.

Tucker, W. and Koren, M. (1971) Ap. J. 168, 283.

Van Regemorter, H. (1962) Ap. J. 136, 906.

Van Regemorter, H. (1965) Ann. Rev. Astr. Ap. 3, 71.

Weisheit, J. (1974) Ap. J. 190, 735.

Wentzel, G. (1927) Zeits. f. Phys. 43, 524.

Weymann, R. (1974) Comments Astr. Space Sci.

Weymann, R. (1965) Phys. Fluids 14, 1701.

Wheeler, J. and Lamb, W. (1939) Phys. Rev. 55, 858.

Wiese, W., Smith, M. and Glennon, B. (1966) Atomic Transition Probabilities, Vol. 1 (NSRDS-NBS-4) (Washington: Government Printing Office).

Wiese, W., Smith, M. and Miles, B. (1969) Atomic Transition Probabilities, Vol. 2 (NSRDS-NBS-22) (Washington: Government Printing Office).

Woltjer, L. (1972) Ann. Rev. Astr. Ap. 10, 129.

Zuckerman, B. and Palmer, P. (1974) Ann. Rev. Astr. Ap. 12, 279.